ON THEORIES

ON THEORIES

Logical Empiricism and the Methodology of Modern Physics

WILLIAM DEMOPOULOS

Edited with a Foreword and Afterword by Michael Friedman

HARVARD UNIVERSITY PRESS

Cambridge, Massachusetts & London, England • *2022*

Second printing

Library of Congress Cataloging-in-Publication Data

Names: Demopoulos, William, author. | Friedman, Michael, 1947– editor.
Title: On theories : logical empiricism and the methodology of modern
    physics / William Demopoulos ; edited with a foreword and afterword by
    Michael Friedman.
Description: Cambridge, Massachusetts : Harvard University Press, 2022. |
    Includes bibliographical references and index.
Identifiers: LCCN 2021020465 | ISBN 9780674237575 (cloth)
Subjects: LCSH: Empiricism. | Physics—Philosophy. | Quantum theory. |
    Science—Philosophy.
Classification: LCC B816 .D46 2021 | DDC 171/.2—dc23
LC record available at https://lccn.loc.gov/2021020465

CONTENTS

Editor's Foreword   *vii*

Introduction   1

1   Logical Empiricist and Related Reconstructions of
    Theoretical Knowledge   18

    1.1   The Partial Interpretation Account of Theories   *18*
    1.2   Carnap on Ramsey Sentences and the Explicit Definition
          of Theoretical Terms   *20*
    1.3   A Proposal of David Lewis and Two Theorems of
          John Winnie   *24*
    1.4   Putnam's Model-Theoretic Argument   *32*
    1.5   Ramsey on Russell's Analysis of Matter and the Partial
          Interpretation of Theories   *39*
    1.6   Constructive Empiricism and Partial Interpretation   *53*

2   Molecular Reality   64

    2.1   The Molecular Hypothesis   *64*
    2.2   Molecular Reality and Brownian Motion   *65*

2.3   The Nature and Status of Perrin's "Connecting Link"   *78*

2.4   Perrin's Argument for Molecular Reality   *83*

2.5   Thomson and the Constitution of Cathode Rays   *93*

3   Poincaré on the Theories of Modern Physics   100

3.1   Poincaré on "True Relations"   *100*

3.2   Robustness versus Consilience   *110*

3.3   Poincaré and Scientific Realism   *113*

3.4   Russell and Poincaré   *116*

4   Quantum Reality   121

4.1   Bohr on the Primacy of Classical Concepts   *121*

4.2   Complementarity, Completeness, and Einstein's Local Realism   *139*

4.3   Bell's Theorem and Einstein's Local Realism   *154*

4.4   Quantum Mechanics and Reality   *172*

Editor's Afterword   187

Notes   205

Bibliography   225

Acknowledgments   235

Index   239

EDITOR'S FOREWORD

William ("Bill") Demopoulos died on May 29, 2017, after an approximately ten-year battle with lymphoma, leaving behind the manuscript *On Theories*. The argument of the manuscript was essentially complete, although the text needed some obvious editorial revisions, especially in connection with the Bibliography. More seriously, however, Demopoulos felt that a final chapter was needed which would perspicuously explain the relationship between the fourth chapter on quantum reality and the earlier chapters, particularly the previous two chapters covering molecular reality and Poincaré's reactions to it. Unfortunately, there was not enough time left, at the end, for Demopoulos to make much progress on this final chapter, which would have served as the counterpart of his Introduction (formerly entitled "Overview of This Study") at the beginning. So, when time was clearly growing shorter, Demopoulos and I agreed that, if needed, I would do my best to prepare the manuscript for publication in accordance with his wishes and see it through to press.

All of Demopoulos's close friends agreed that the last decade of his life was extraordinarily productive. It culminated, in particular, with the collection of essays *Logicism and Its Philosophical Legacy* (2013) and the present book, *On Theories,* which, for its part, represents the culmination of his work in the philosophy of physical science. During this period, Demopoulos and I spoke at length by telephone almost every weekend. And it was from this experience, together with regular visits to London, Ontario, that I derived my appreciation of what he was attempting to do in *On Theories*. As I was considering what I could reasonably accomplish in the wake of his death, however, it quickly became clear that I was in no position to write the final chapter that Demopoulos might have written had he lived. He did not leave behind even a short draft of this chapter, and, more importantly, my own philosophical style is very different from his. What I could reasonably attempt, first, was to locate Demopoulos's argument against the background of his more recent work in the philosophy of science (as represented in *Logicism and Its Philosophical Legacy*) and, second, by locating his discussion of quantum reality against the background of his ongoing work on the interpretation of quantum mechanics throughout his career. The present Foreword is devoted to the first; the concluding Afterword to the second.

Logicism and Its Philosophical Legacy, not surprisingly, focuses centrally on views of the relationship between logic and mathematics in the classical logicist tradition that begins with Frege and continues with Russell, Ramsey, and Carnap. Two of that book's essays, however, "Carnap's Thesis" (chapter 2) and "On Extending 'Empiricism, Semantics, and Ontology' to the Realism-Instrumentalism Controversy" (chapter 3), introduce new material concerning the relationship between a priori mathematical and empirical physical theories that is especially relevant to the argument of *On Theories.*

"Carnap's Thesis" proposes a novel understanding of Carnap's distinction between a priori and empirical theories drawing on Demopoulos's earlier work on neo-logicism and Hume's Principle, as discussed in Demopoulos's important edited volume *Frege's Philosophy of Mathematics* (1995). Already here Demopoulos had suggested that Hume's Principle is not best understood as an analytic truth but rather as a *criterion of identity,* which, in good Fregean style, can be seen as an *analysis* of the concept of natural number but not as a reductive *analytic definition* of this concept within pure logic; this, in the end, is the lesson of the failure of Axiom V of the *Grundgesetze*. The resulting novel understanding of

Carnap's distinction then follows as a kind of corollary. The criterion of identity provided by Hume's Principle applies equally to applied as well as pure arithmetic, insofar as any sortal concept whatsoever, and, in particular, any empirical sortal concept (such as "people in this room") figures in an account of the application of the concept of number (via Hume's Principle) to any appropriate empirical domain. By contrast, criteria of identity for empirical concepts in physics, such as temporal simultaneity and physical spatial congruence, are constrained by clearly empirical rather than purely logical factors. Temporal simultaneity, for example, is determined by empirical methods of measurement (using light signals and the like), and so, of course, is physical spatial congruence (as in Einstein's appeal to practically rigid bodies).

Here we find the germ of the asymmetry Demopoulos emphasizes between physical theories and logico-mathematical theories in the work of Carnap. But the really decisive asymmetry, from the point of view of *On Theories,* emerges in the following chapter 3 of "On Extending 'Empiricism, Semantics, and Ontology' to the Realism-Instrumentalism Controversy" (Demopoulos 2013) (which had previously appeared in the *Journal of Philosophy* in 2011). Demopoulos is here concerned, among other things, with replying to an objection to Carnap's paper raised by Penelope Maddy in *Second Philosophy: A Naturalistic Method* (2007). The objection is that Carnap cannot provide a reasonable interpretation of the question of whether atoms and molecules exist, since we begin with a simpler language for recording scientific observations (say, the "thing language"), and we must then extend this language to contain theoretical terms (say, "atom" or "electron") before we can even raise the question of the existence of such things. But, according to Carnap (Maddy's objection goes), all choices of language or linguistic framework must be purely pragmatic or conventional, which is precisely the mark of an *external* rather than internal question. This (the objection concludes) flies in the face of the fact that the surprising evidence marshalled by Einstein and Jean Perrin based on the phenomenon of Brownian motion provides a perfectly good *internal* answer to the question of (atomic and molecular) existence using only our standard methods of inquiry.

It appears, however, that this is by no means a powerful objection to Carnap's conception of linguistic frameworks. The concepts of atom and molecule, understood in something like our modern usage, have been part of scientific language since at least the beginning of the nineteenth century. Moreover, increasingly sophisticated theories of these entities were developed and applied throughout the nineteenth century. Nevertheless, there was no scientific consensus about their existence or reality until the early years of the twentieth century, on the basis of delicate new experiments on Brownian motion drawing on the work of Einstein, Perrin, and others. And it is especially relevant to the argument of *On Theories,* in particular, that the very sophisticated late-nineteenth- and early-twentieth-century work on the kinetic theory of gases by Maxwell and Boltzmann yielded theories of the behavior of atoms and molecules providing rigorous hypothetico-deductive explanations of observable phenomena. But such explanations were standardly taken to be less than completely convincing with regard to precisely the question of existence or reality. This is because a merely hypothetical explanation of observable phenomena leaves it entirely open as to whether an alternative hypothesis (in this case, one not postulating atoms and molecules) can provide an equally good or even superior such explanation.

From this point of view, Maddy's objection that a new language containing the concepts of atom and molecule must first be chosen on pragmatic grounds, and that such a choice of language must always concern an external question for Carnap, turns out to be a red herring. Carnap would have no difficulty in appealing to our standard methods for answering such questions on purely internal scientific grounds—and, indeed, it appears that precisely this is what happened in the case of the surprising new experimental evidence based on Brownian motion. I shall come back to the special character of this kind of evidence and the nature of its demonstrative power below. But I first want to note that Demopoulos himself, in his very detailed and subtle examination of the application of "Empiricism, Semantics, and Ontology" (Carnap 1950) to the realism-instrumentalism controversy, calls attention to problematic features of Carnap's overall position in this philosophical controversy.

What is most problematic, Demopoulos explains, is Carnap's use of the Ramsey-sentence reconstruction as a full and adequate representation of the factual content of a physical theory. Although Carnap's explicit use of this representation occurs after the publication of "Empiricism, Semantics, and Ontology" (1950), elements of his earlier treatments of such theories (going back at least to Carnap 1939) clearly anticipate Carnap's later Ramsey-sentence reconstruction. The crucial point is that the Ramsey-sentence reconstruction divides the vocabulary of a theory into two exclusive categories, observational and theoretical, and then applies existential generalization to all the theoretical terms. The result is a representation that captures only the logical structure of the theoretical part, which then draws its factual content from its deductive relations to the observational part. Carnap takes this representation to be neutral between realism and instrumentalism, because, after all, it does claim existence for objects in the domain over which the quantifiers in the theoretical part range. What is then problematic, however, is that the objects whose existence is asserted might just as well be purely mathematical objects (numbers, sets of numbers, and so on), and we of course already know that such objects exist on purely logico-mathematical grounds. Leaving this problem aside, moreover, the Ramsey-sentence reconstruction takes the meaning of the theoretical part to be effectively specified only by the sentences or laws of the theory as a whole (including both theoretical and observational terms), so that any change to these laws, no matter how small, must change the meanings of the theoretical terms—and, indeed, of all of them.

In both *Logicism and Its Philosophical Legacy* and *On Theories,* Demopoulos finds a fundamental problem with the way in which the logical empiricists (including Carnap, of course) model their philosophy of physical science on Hilbert's axiomatic conception of purely mathematical theories. The Hilbertian conception of the wide generality of mathematical theories rests on the view that mathematical vocabulary is abstract, reflecting only the logical structure provided by the axiomatization in question. It is in this sense (and only this sense) that the mathematical vocabulary is "uninterpreted." According to the logical empiricists, however, while the theoretical vocabulary of a physical theory is similarly uninterpreted, such a theory must also possess an observational vocabulary with a fixed interpretation. It is this (and only this) that distinguishes physical theories from abstract mathematical theories and, accordingly, makes them "*partially* interpreted." And what provides a partially interpreted theory with an empirical content absent from purely mathematical theories are mixed sentences containing both theoretical and observational terms, such that definite observational predictions are then possible.

While both *Logicism and Its Philosophical Legacy* and *On Theories* challenge the idea of Hilbert's axiomatic conception of purely mathematical theories as a model for empirical physical theories, the two offer importantly different diagnoses of the underlying problem. The former focuses on what Demopoulos calls the *structuralist thesis,* the view that specifically theoretical terms of a physical theory are constrained only by their logical structure (see especially chapters 4 and 7 of the *Logicism* volume). The clearest case of this situation arises in Carnap's use of the Ramsey-sentence representation, which, Carnap thinks, exhaustively represents the empirical content of the theory in question. The Ramsey sentence, however, existentially generalizes over all the theoretical terms, so that it is *true* in a model if and only if the matrix of the theory—which results from replacing the original theoretical terms by instantial variables—is *satisfiable* in the model. The observational terms of the theory, by contrast, are constant terms with fixed interpretations, so that this part of the theory is true (or false) in the ordinary sense. Or, to put the point in a slightly different way, the theory as a whole is true if and only if it is empirically adequate—that is, if and only if the observational part of the theory is true in the ordinary sense. The situation is in this sense analogous to Hilbert's conception of purely mathematical theories, which can be understood as substituting consistency (or satisfiability) for truth in this case. However, while this is relatively unproblematic (albeit not completely unproblematic) in the mathematical case, it appears to be highly counterintuitive in the case of empirical physical theories.

At the beginning of the Introduction to *On Theories,* Demopoulos draws attention to rather different problems with the Hilbertian model— epistemic and methodological problems rather than semantic ones. Thus, after suggesting once again that overemphasis on the logical and mathematical aspects of Hilbert's conception can be philosophically problematic for our understanding of empirical physical theories, Demopoulos rejects the "method of hypotheses"—which also includes inference to the best explanation. This method is problematic, Demopoulos continues, because "[it] misses altogether the phenomenon of 'theory-mediated measurement' and the special methodological role such measurement plays in securing the empirical determination of theoretical parameters," and he also explains (quite rightly) that the recent "interest in theory-mediated measurement is largely the result of a reevaluation of Newton's importance as a methodologist" (see endnote 1 of the Introduction, which highlights George Smith's work on Newton). It is well worth our while, therefore, to elaborate further on Newton's distinctive methodology, for his example is what first reveals the special character of this kind of evidential support and the nature of its demonstrative power.

Newton's distinctive methodology is delineated explicitly in the four Rules for the Study of Natural Philosophy formulated at the beginning of Book 3 of the *Principia.* The most distinctive of these are the last two, which, in the translation of I. B. Cohen and A. Whitman (1999, pp. 795–796), read as follows:

Rule 3: *Those qualities of bodies that cannot be intended and remitted [i.e., qualities that cannot be increased and diminished] and that belong to all bodies on which experiments can be made should be taken as qualities of all bodies universally.*

. . . .

Rule 4: *In experimental philosophy propositions gathered from phenomena by induction should be considered either exactly or very
nearly true notwithstanding any contrary hypotheses, until yet other
phenomena make such propositions either more exact or liable to
exceptions.*

This rule should be followed so that arguments based on induction
may not be nullified by hypotheses.

Rule 3 is therefore a principle of universal induction, Rule 4 a principle
for avoiding the interference of conjectural hypotheses in the inductions
featured in Rule 3. In order to grasp the full import of these two rules,
however, we need to take account of the contemporaneous context.

Rule 3 was added in the second edition of the *Principia* (1713) to
address the opposition to Newton's theory of universal gravitation
expressed by his contemporaries Huygens and Leibniz, while Rule 4,
added to the third edition (1726), served principally to reinforce the Newtonian defense in light of continued such opposition. Huygens and
Leibniz rejected the idea that gravity was a nonmechanical force, oper-
ating by action at a distance rather than action by contact, and they
proposed instead an aetherial vortex theory that could, in principle,
explain the celestial orbital motions in a way that was mechanically acceptable. But this theory, at the time, was merely hypothetical or conjectural, because there was no practicable way to measure the unobservable motions in the aetherial vortex so as to provide an empirical basis
for deriving the observable celestial motions from the proposed vortex
theory. The point is that Newton's law of universal gravitation, despite
the fact that it failed to offer a mechanical explanation, was grounded
in a proper empirical argument by induction from the phenomena,
while the vortex theory, at least at the time, was not even close to such
a grounding.

The Newtonian conception of the relationship between induction from the phenomena, on the one side, and avoiding contrary evidence based on merely conjectural hypotheses, on the other, is richer and more intricate than it may first appear. The Newtonian conception of impressed force, for example, is that of a certain definite *physical quantity:* the quantitative increase or decrease in the state of motion of a body (its acceleration) in proportion to the mass or quantity of matter of the body in question. Both of the other two physical quantities—*true* acceleration (increase or decrease in the state of motion relative to an inertial frame) and mass or quantity of matter—are theoretical rather than observational quantities. Nevertheless, Newton's argument for the law of universal gravitation in Book 3 establishes theory-mediated measurements derived from phenomena for both. Thus, if we consider the relative motion of the moon toward the Earth and the similarly relative motion of Venus toward the Sun, we then find—in a way that does not beg the question of whether the Earth or the Sun is the center of motion of the solar system—that the acceleration of Venus toward the Sun (at a given distance) is much, much larger than that of the moon toward the Earth (at the same distance). But it then follows that the quantity of matter (active gravitational mass) of the Sun is much, much greater than that of the Earth. Therefore, the center of gravity or center of motion of the Earth-Sun system is almost (but not quite) at the center of the Sun, and the seemingly intractable Copernican problem has now been solved empirically—and thus *internally*—by precisely the theory-mediated measurements in question.

In this example, the central mathematical relationship between observable phenomena (the observable relative motions between primary bodies and their satellites in the solar system) and the relevant theoretical quantities (gravitational force, true acceleration, and quantity of matter) asserts that the (true) acceleration of a satellite toward its primary body arising from the gravitational force exerted by its primary body is proportional to the active gravitational mass of the primary body and inversely proportional to the square of the distance between them. What is important is that the (true) acceleration of the satellite arising from the gravitational force of its primary body depends *only* on the other two theoretical quantities—and is independent, therefore, from the mass or other internal properties of the satellite itself. This property—the "acceleration-field" property in Howard Stein's (1967) apt terminology—is unique to gravitational force and therefore opens the way for a non-question-begging determination of all three theoretical quantities. This determination, in turn, proceeds by successive approximations.

In particular, for the systems comprising Jupiter and its moons, Saturn and its moons, and the Sun in relation to the five uncontroversial planets (Mercury, Venus, Mars, Jupiter, and Saturn), observable deviations from the strict inverse-square acceleration-field property due to perturbations by bodies other than the primary body were, at the time, quite negligible. The one exception was the motion of the Earth's moon relative to its primary body, which exhibited observable (although still very small) deviations from the inverse-square proportion. Newton, in the *Principia*, was aware of this, and he (rightly) took these deviations to be due to the Sun. However, he was not able fully to calculate this effect of the Sun, which was only shown to follow from the inverse-square proportion by Euler and Alexis Clairault in the 1740s. Nevertheless, if Newton's preliminary solution to the three-body problem did not survive, his methodology for articulating and refining such problems within the theory of universal gravitation enjoyed a considerably longer life.

Newton's procedure of successive approximations is essential for the move from purely inductive relations among observable phenomena to theory-mediated inferences involving the theoretical quantities of gravitational force, (true) acceleration, and quantity of matter. Newton begins, in fact, with what he himself calls "Phenomena" at the beginning of Book 3 of the *Principia:* such Phenomena are confined to geometrical, kinematical, and optical features of the orbits (where the phases of Mercury and Venus are paradigmatic of the latter), while the three theoretical quantities in question all involve, in addition, causal-dynamical features. Thus, the Phenomena at the beginning of Book 3 proceed, in the first instance, from Kepler's three laws of planetary motion, now taken to apply to the systems of satellites and primary bodies discussed above. In deriving the inverse-square acceleration-field property for all these systems (with the exception of the Earth-moon system), Newton appeals to the spatiotemporal (geometrical-kinematical) features of the relative observable motions in accordance with Kepler's three laws. And, when small deviations do appear, as we have seen, Newton than looks for a further gravitational source (also observable) capable of making the inference, in the words of Rule 4, "*either more exact or liable to exceptions.*" Clearly, if we allow the interference of any "*contrary hypotheses*" not yet derived from "*phenomena*" in the same approximative sense as Rule 4, we open the door for merely conjectural reasoning, as in the vortex theory.

In sum, inferences involving theory-mediated measurements, as deployed by Newton, involve all three of the elements noted above. Observable, nondynamical Phenomena governing only relative motions form the basis for the inference in question. The causal-dynamical quantities of impressed force, true acceleration, and quantity of matter, as mathematically structured by the Axioms or Laws of Motion preceding Book 1, form the target of the inference—these three Axioms are what "mediate" the inference. And, finally, the inference itself involves a potentially infinite sequence of successive approximations as we iterate the process, in accordance with Rule 4, in a chain of such inferences resulting in what we hope will be ever more exact determinations of the causal-dynamical theoretical quantities and their emerging inductively established relations. The distinctive (inverse-square) acceleration-field property of specifically gravitational force is perhaps the most important of these relations, so let us look at this example in more detail.

The (inverse-square) acceleration-field property is a direct inductive extrapolation in accordance with Newton's Rules 3 and 4. The induction begins in the work of Galileo as a constant acceleration for all falling bodies near the surface of the Earth (regardless of their intrinsic properties) and continues in the work of Kepler as his third or harmonic law for the five primary planets and then (in the hands of contemporary astronomers and Newton himself) the satellites of Jupiter and Saturn. The harmonic law implies that the inverse-square proportion continues to hold from orbit to orbit in any multiple satellite system as a (variable) function of distance. It thereby generalizes and corrects Galileo's original law, which only holds (to a very high degree of approximation) for bodies near the surface of the Earth. Newton shows this in the so-called moon test near the beginning of Book 3, which extends the Galilean law to an inverse-square proportionality governing both the (now almost) constant accelerations for bodies falling near the surface of the Earth toward its center and the variable acceleration of the moon as we take it to be falling (accelerating) toward this same center. The gravitational acceleration of the moon toward the center of the Earth exhibits the inverse-square acceleration-field property all the way down from its actual distance to the surface in a direct line to the center of the Earth, where the acceleration in question (at the surface) is then identical to the Galilean constant (g) to an extremely high degree of approximation. We also know, however, that there is yet another important complication here, for the moon also betrays a further (albeit much smaller) acceleration toward the Sun, which must be separated from the first acceleration by perturbational methods. And, as we have seen, this leads to the first successful development of such methods after Newton, beginning with the work of Euler and Clairault in the 1740s on the motion of the moon.

Our direct inductive extrapolation includes a further important property of gravitational acceleration fields. The fields of different bodies in the solar system vary markedly in their strengths, insofar as they involve very different accelerations generated at any given distance: The field of the Sun is much stronger than that of the Earth and, more generally, than that of any other body in the system; Jupiter's acceleration field is significantly stronger than those of all the other planets; and so on. This means that the fields of gravitational force surrounding each body vary, from body to body, in direct proportion to the accelerations thereby generated at any given distance; if we fix the gravitating body in question, the resulting accelerations vary only with the distance from that body in accordance with the inverse-square law. Every gravitating body, therefore, is associated with a distinct measurable physical quantity of (absolute) gravitational field strength, which we now call "active gravitational mass." This direct inductive extrapolation suffices, by itself, to complete the Newtonian resolution of the dispute between geocentrism and heliocentrism: The center of active gravitational mass in the solar system (i.e., the center of accelerated motion in the system) lies always very close to the center of the Sun.

Returning now to the development of perturbational methods, the most important contributions after those of Euler and Clairault, continuing with the motion of the moon and then attacking interplanetary perturbations as well, were those of d'Alembert, Lagrange, and Laplace. Laplace, in particular, first provided an adequate understanding of the so-called great inequality of Jupiter and Saturn—as a consequence, once again, of the Newtonian law of universal gravitation. In 1781 William Herschel telescopically observed a new planet beyond Saturn, Uranus, whose perturbative interactions with both Saturn and Jupiter could also be studied with the help of Newton's law and Laplace's analytical methods. It turned out, however, that what was most important about the discovery of Uranus was the way in which it then led to the discovery of Neptune— first as a hypothetical prediction and later as a (telescopically) observed reality.

The story is relatively familiar. In the years 1845–1846, John Couch Adams and Urbain-Jean-Joseph Leverrier independently calculated where a more distant planet might lie, and what kind of orbital properties it might have, in order to explain the residual perturbative effects in the orbit of Uranus left over from the already known such effects by Saturn and Jupiter. In order to obtain quantitative predictions, however, each of the two astronomers needed to hypothesize a mass for the new planet, together with a size and eccentricity for its orbit. They did the best that they could on the basis of the telescopic observations of Uranus available to them, with the result that they each (independently) predicted a heliocentric longitude of the new planet for September 23, 1846, between 326 (Leverrier) and a bit more than 329 degrees (Adams). Johann Galle of the Royal Observatory in Berlin, at the request of Leverrier, searched for and observed the new planet (very near the ecliptic) at just short of 327 degrees of heliocentric longitude, and the real planet Neptune was thereby finally detected observationally.

In order to confirm that Neptune was indeed the gravitational source of the residual perturbations in question, we still needed to know, independently, what the value of its active gravitational mass actually was—at least to a high degree of approximation. So it was very fortunate, in this case, that seventeen days after Galle made the first reliable telescopic observations of Neptune in 1846, William Lassell observed its largest and nearest satellite, which was later named Triton. Multiple observations of Triton's orbit then yielded the first values of Neptune's active gravitational mass of the same order of magnitude as modern values. In 1898, moreover, on the basis of years of accumulating more accurate estimations of the distance between the two planets and the shape of Neptune's orbit, Simon Newcomb was then able to determine the active gravitational mass of Neptune—and therefore the source of its gravitational acceleration field—with a value that became standard for the next ninety years (until the Neptune flyby of Voyager 2 in 1989).

The discovery of Neptune represented a spectacular success of the Newtonian gravitational law. Eventually, however, there was an almost equally spectacular failure of this law—insofar as it finally led to the radically revised theory of gravitation of Einsteinian general relativity. The problem with the Newtonian law, in particular, involved a continuing failure to find an appropriate Newtonian gravitational source for the very small residual perturbations in the motion of the perihelion of Mercury. Simon Newcomb, moreover, had extensively investigated this same anomalous motion in 1895, and he thus played a central role in both the success of Newton's theory in accounting for the perturbations of Uranus and its failure exhaustively to account for the perturbations of Mercury.

But the point I most want to emphasize is that this failure of the Newtonian gravitational law involved another spectacular success of the Newtonian empirical method. Of great importance here, moreover, is the fact that the anomalous motion of Mercury's perihelion is a *residual:* a phenomenon left over from the previously known perturbations that had been successfully derived from the Newtonian gravitational law. It would not only have been completely impossible to "observe" this remaining anomaly independently of the Newtonian law, but we would also have been left without any useful guidance for how to proceed—namely, to envision and then find empirical evidence for a radically different kind of gravitational source (relativistic space-time curvature). We can therefore make a stronger point. Not only did the Newtonian method prove its worth once again, but the Newtonian gravitational law continued to prove its worth at the same time. For it was only on the basis of this same law that we were then led to a better theory capable of accounting for the very small residual phenomena remaining in the Newtonian theory.

Newtonian theories like that of universal gravitation aim to address the continual accumulation of empirical evidence over time—in such a way that they can, and typically do, eventually lead to better theories. The goal is not to find the completed final theory, for this could only be a theory correct to any arbitrary degree of approximation. On the contrary, the point of Newtonian methodology is to exploit a series of successive approximations taken always—in accordance with Rule 4—to be capable of further refinements. Overarching general theories like the law of gravitation, for Newton, can therefore be viewed as essential instruments for discovering new and better approximations—which successively take their place in the evolving scientific system as well-established (but still revisable) results. Among such results, in particular, we find the Newtonian resolution of the ancient dispute between geocentric and heliocentric astronomy in terms of the center of gravity of the solar system, the discovery of the planet Neptune within this same Newtonian theory, and (most dramatically) the further discovery of the precise limits of Newtonian theory within Einsteinian gravitational theory. That Newtonian theory-mediated measurements can and did result in such a change in the gravitational law shows that the "theory" in question need not include the gravitational law itself.

When we speak of Newtonian theory-mediated measurements, therefore, we do not understand the concept of "theory" in the logical sense that has dominated recent philosophy of science—as either a fixed and completed set of sentences or a fixed and completed class of models. As applied to what we now call the "Newtonian theory of universal gravitation," for example, it would include both the three Laws of Motion and the law of universal gravitation. If we take Newton's gravitational theory to be fixed and complete in this sense, it seems that Newtonian theory-mediated measurements must already *presuppose* the completed theory of universal gravitation. And it then becomes extremely difficult to see how Newton could have applied such measurements to establish his theory of gravitation in the first place. If, on the other hand, the measurements in question presuppose only the three Laws of Motion, for example, together with the Newtonian Rules for the Study of Natural Philosophy, this difficulty, at any rate, need not arise.

Demopoulos considers the scope of the "theory" in theory-mediated measurements in Chapter 2 of *On Theories* (entitled "Molecular Reality"). Section 2.4 on "Perrin's Argument for Molecular Reality" directly confronts a conception of theory-mediated measurement that is decidedly *non*-Newtonian and must therefore be sharply distinguished from the Newtonian conception. In particular, this section considers Bas van Fraassen's (2009) rebuttal to Perrin's argument for molecular reality drawing on Clark Glymour's (1980) "bootstrapping" account of theory confirmation. The rebuttal, briefly, is that it is misleading to take Perrin's experimental investigations of Brownian motion to be evidence for the existence or reality of molecules—since these investigations, according to van Fraassen, are carried out wholly within Maxwell-Boltzmann molecular theory and therefore presuppose the existence of molecules. As Demopoulos makes clear, however, these investigations of Brownian motion were taking place within a historical context of serious uncertainty surrounding the Maxwell-Boltzmann theory based on emerging new phenomena in the science of specific heats, for example, indicating a failure of the equipartition of energy. Perrin therefore took care, against the background of such phenomena, not to rely on the Maxwell-Boltzmann theory outside of its already established domain of successful application.

Far from presupposing this theory as a fixed and established edifice, Perrin's argument for molecular reality began with the behavior of the microscopically observable Brownian particles themselves and drew conclusions about the (then) unobservable molecules by inductive extrapolations from the behavior of the Brownian particles. Moreover, Perrin left the detailed description of the fundamental properties of the molecules—their sizes, shapes, and structures—largely open. Aside from being a manifold of discrete elements rather than a continuum, where the elements in question are taken to be of relatively small size, any more detailed description of their structural properties needed to be consistent with the gradually emerging quantum theory. It is no wonder, then, that Perrin's work represented only the tip of the iceberg in modern atomic-molecular theory, whose further empirical investigation, beginning with the work of Bohr on atomic and molecular structure, required more delicate and sensitive experiments. These experiments, in particular, involved electromagnetic (rather than thermal and kinetic) interactions between the (then) unobservable electrons, protons, and neutrons hypothesized in Bohr's theory and the observable bands of emitted radiation that could then be spectrographically studied.

Two points are particularly important here. First, Perrin did not advance the theory of atomic and molecular structure, a task that was then reserved for Bohr. Second, the experiments Bohr made required a new piece of technology, the spectroscope, which was not available to Perrin. Perrin had neither the Bohrian discoveries in atomic structure nor the new instruments for contributing to them. The case of Newton and gravitational theory, by contrast, involves both a robust theoretical apparatus and an explicit general methodology—Newton's Rules for the Study of Natural Philosophy (especially Rules 3 and 4)—for further developing the theory in question. And, as it turns out, Newton's methodology can and does support a radically new theory, the general theory of relativity, as a replacement for Newton's law of gravitation. *On Theories* is where William Demopoulos fully responds to the overemphasis in the philosophy of physics on mathematical theories and argues strongly for Newtonian experimental methodology.

Introduction

Theories—by which I mean the theories of [modern]* physics—are the instruments with which we seek to represent the structure and constitution of reality. Beginning from this premise, it is clear that any discussion of theories must illuminate the status of the abstract principles to which physics appeals when it proposes general constraints on physical processes, and it must illuminate the status of the existence claims to which physics is led in its attempt to uncover the nature of matter. In addition to an examination of the character of the theoretical claims that are peculiar to modern physics, a discussion of theories should also contribute to our understanding of the basis for our confidence in whatever truth we take such claims to possess.

The early twentieth-century experimental success of atomism established the appeal to unobservable entities as a permanent feature of physical theory, and this raised the question of the nature of the support that attaches to claims purporting to be about entities that transcend observation. In the case of the logical empiricists, the search for an account of theoretical claims which appeal to unobservable entities culminated in the "partial

*Editor's note: I shall return later to the meaning of characteristically "modern" physics.

interpretation" view of theories, one of whose principal goals was to address the prima facie challenge to empiricism such claims represent by clarifying their empirical status.

The partial interpretation account of theories is arguably the first systematic development of a theory of theories. There were many influences which shaped this account of theoretical knowledge, but one of the most striking, and the one on which I will focus, derived from the perspective afforded by modern logic and Hilbert's work on the axiomatic method. The partial interpretation account divides the vocabulary of the language of physics into "theoretical" and "observational" terms and reconstructs the theoretical claims of physics by sentences whose nonlogical vocabulary is exclusively "theoretical." The account assimilates its reconstruction of the theoretical claims of physics to Hilbert's analysis of the statements of a purely mathematical theory. It recovers the empirical character of physics from the fact that, on this reconstruction, theoretical claims, by contrast with the claims of a purely mathematical theory, are related to an "observation language" consisting of sentences whose nonlogical vocabulary is "observational." The relation between theoretical and observation vocabulary, and derivatively, the relation between theoretical and observation sentences, is effected by sentences whose nonlogical terms are both theoretical and observational; these are the mixed sentences or "correspondence rules" of the partial interpretation account. The interpretation is "partial" because only the observation terms are completely understood, and only they are interpreted.

I will argue that the partial interpretation reconstruction of theories that purport to be about entities which transcend observation is based on a fundamental misconception regarding the character of the theoretical claims these theories express. On the partial interpretation reconstruction, the ascription of truth to theoretical claims fails to capture the fact that such claims, if true, express salient truths about reality. This is an unintended consequence of the partial interpretation reconstruction of theoretical claims. Its reconstruction of ascriptions of truth to observation sentences does not exhibit this feature, but it stands in sharp contrast with the approach the reconstruction is committed to in the case of theoretical claims. We will see that this misconception stems from an incorrect assessment of the epistemic warrant that theoretical claims involving unobservables enjoy.

Aside from the historical interest of a study of the logical empiricist theory of theories, what can be gained by revisiting an approach which, like the partial interpretation view of theories, is generally thought to have been superseded by subsequent developments? There would be very little to be gained, were it not for the fact that the nature of its incorrect assessment of the methodological basis for theoretical claims about unobservable entities has gone largely unrecognized and persists even in views that have sought to replace it. In particular, the notion that the difficulty with logical empiricism derives from its overly "syntactic" conception of theories not only fails as a fair representation of the partial interpretation view of theoretical knowledge, it also deflects attention from the real difficulties that confront the view. In particular, the extent to which the inadequacy of the logical empiricists' assessment of the evidential basis for theoretical existence claims rests on a reliance on the "method of hypothesis"— hypothetico-deductive reasoning and reasoning by inference to the best explanation—has been insufficiently appreciated. It is surprisingly easy to show that many of the alternative assessments that have been advanced by logical empiricism's principal critics share the central inadequacy of the view they reject.

The first part of this study traces the development and refinement of the partial interpretation view of theories in the work of Ramsey and Carnap. I also describe the bearing of a contribution of David Lewis on Carnap's mature formulation of his view. I then turn to the exposition of internal difficulties with the partial interpretation account that were raised by John Winnie. The presentation of these difficulties rests on considerations of an abstract and logical nature, as does the problem which I argue is posed by Hilary Putnam's "model-theoretic argument." While it may be possible for a proponent of partial interpretation to evade the issues raised by Winnie, the difficulty which I argue can be extracted from the model-theoretic argument shows that a different approach to theoretical knowledge is required. But although the logical investigation of this part of the discussion suffices to isolate the fundamental misconception in the partial interpretation reconstruction of theoretical claims, it is too abstract to point the way to a better account of the basis for according such claims empirical status, or to a more accurate representation of their empirical support. To make progress in these directions, it is necessary to delve more deeply into the sources of the doctrine of partial interpretation and to consider particular developments in physics that are of central importance to an adequate account of theoretical knowledge.

The partial interpretation account of the empirical basis of theoretical claims involving unobservable entities supports the logical empiricist conception of the nature of the evidence that attaches to theories of molecular, atomic, and subatomic reality. The method of hypothesis is, in its turn, naturally suggested by the partial interpretation account of how theories acquire their empirical status and the idea that the status of claims which concern theoretical parameters is adequately addressed by the inclusion of statements that contain theoretical and observational vocabulary. We will see that the partial interpretation account captures only very incompletely what actually underlies our judgment that a parameter which applies to unobservable entities is empirically well founded. In particular, the method of hypothesis misses altogether the phenomenon of "theory-mediated measurement" and the special methodological role such measurement plays in securing the empirical determination of theoretical parameters.

The methodology of theory-mediated measurement is not new. But its careful study is. The interest in theory-mediated measurement is largely the result of a reevaluation of Newton's importance as a methodologist.[1] A novel feature of the theory-mediated measurement of a hitherto undetermined parameter P, and one which distinguishes it from the case of a successful prediction of an observation statement, is the manner in which the application of the theoretical framework both guides, and is itself guided by, experimental design. A theory-mediated measurement of P exploits P's functional relationships with other parameters, parameters for which there exist experimental techniques for determining their values, to infer a value for P. In some instances, the parameters that fulfill this role are introduced for the specific purpose of measuring P. An illustrative example which is relevant to our concerns is given by John Townsend's introduction of the mobility of gaseous ions and their coefficient of diffusion. These parameters were identified by Townsend, who also devised the first experimental techniques for measuring them. He showed by a theoretical analysis how the quantity ne, which is the product of the number n of molecules in a specified volume of air (at a prescribed temperature and pressure) and the average charge e on an atmospheric ion, functionally depends on the mobility of gaseous ions and their coefficient of diffusion. From the empirical determination of the value of ne and knowledge of n, it is possible to obtain an empirical determination of e. Townsend's discovery facilitated comparison of ne with its determination by methods in which the significance of n is the same, but e is the average charge on a univalent ion in electrolysis. This led, ultimately, to the discovery that the charge on gaseous and univalent ions is the same, and this in turn figured essentially in an extended argument that concluded with the discovery of the complex structure of the atom.[2]

The functional relations on which theory-mediated measurements rest are typically general and independent of the existential hypothesis which their application supports. A case in point is the use of Stokes's Law, which relates the rate of fall of a spherical object to its density and the density and viscosity of the medium in which it is immersed. This law was used by J. J. Thomson and his Cavendish collaborators to infer the size of a spherical water drop from its velocity. The volume of the water drop is of interest, because it is indicative of the *charge* on the ions it contains under certain experimentally controlled conditions. Stokes's Law is of course a piece of fluid mechanics, unrelated to the electrons whose properties Thomson sought to determine. The use of the law in such calculations was not to obtain, as an instance of it, the prediction of a velocity, but the measure of a parameter which when combined with an intricate piece of reasoning and experimental design, led to the determination of a property of electrons. Thus a novel methodological contribution of the consideration of theory-mediated measurement is the shift in emphasis it induces from the *prediction* of a parameter value, as a means of confirming a hypothesis, to the parameter's *measurement* under various assumptions involving its relations to other parameters. The greater the variety of independently determinable parameters and functional relationships that yield the same value for a particular constant or quantity, the more *robust* is its theoretically mediated determination. Robustness is further enhanced when the parameters and functional dependencies belong to different branches of physics and thus arise in different theoretical frameworks.

Townsend, Thomson, and others carried the analysis of molecular reality into the domain of the subatomic. But their analysis occurred in a theoretical context in which the existence of both the molecular and atomic levels—if not that of a subatomic level—was taken for granted. This points to why the use of the methodology of theory-mediated measurement that Jean Perrin exploited in his attempt to secure molecular reality is of particular philosophical interest: Perrin did not assume the reality of molecules but sought to establish our epistemic access to the observation-transcendent or "theoretical" domain they inhabit on the basis of our experience with the observable phenomenon of Brownian motion. Theory-mediated measurement plays an essential role in Perrin's argument to this conclusion, as does the robustness of his determination of various parameters, perhaps most prominently, Avogadro's constant. Indeed, the robustness of the determination of Avogadro's constant is a central premise in support of Perrin's interpretation of the constant as the number of molecules in a mole.[3] But because Perrin was engaged in undertaking the very first steps toward establishing our epistemic access to a domain of observation-transcendent entities, his arguments possess a character and subtlety that have a special interest for our study.

The logical empiricists' conception of evidential support makes it difficult to distinguish the methodological basis of atomism, as it was developed by the seventeenth-century corpuscularian philosophers, from its nineteenth-century articulation in the molecular-kinetic theory of heat, or from its subsequent refinement in light of the experimental and theoretical advances of the early twentieth century.[4] The difficulty arises from the fact that the methodological basis for claims about the constitution of matter that appeal to unobservable entities consists of more than their correct prediction of observation statements. In the cases of greatest interest, it also includes diverse but concordant theory-mediated measurements of the parameters that qualify these entities. Perrin's articulation of a framework of assumptions and experimental techniques sufficient to enable concordant measurements of great accuracy and increasing precision proved decisive in establishing molecular reality. Both the design of such measurements and Perrin's deployment of them in support of molecular reality involved an interplay between theory and experiment that is not easily captured within the logical empiricist reconstruction of theories. As a result, the evidentiary importance of theory-mediated measurement was largely missed by the reconstruction's proponents.

The logical empiricists acknowledged that when empirical laws are explained within a new theoretical framework, they invariably occur transformed by "corrections from above"—corrections mandated by the theoretical principles of the framework into which they have been incorporated. But this concession to the role of theory in shaping our conception of evidence did not challenge the primacy and independence from theory which logical empiricism accorded the observation language of its reconstruction. Nor did it challenge the logical empiricists' conception of the evidentiary framework of physics. That conception is built on a very specific model of predictive success and its role in establishing existence claims. While the logical empiricist distinction between what belongs to theory and what to observation might be a suitable guide for representing the notion of evidential support that derives from this model, it is at best a coarse grid with which to capture the contribution of robust theory-mediated measurements.

Given their assessment of the evidential support for theoretical existence claims, it is unsurprising that advocates of partial interpretation supposed themselves justified in representing the realism-instrumentalism controversy as a "practical" question about the use of theoretical vocabulary rather than a "theoretical"— indeed, factual—question about the reality of a collection of unobservable entities.[5] Regarding the question of realism versus instrumentalism, Carnap writes that it "should not be addressed in the form: 'Are theoretical entities real?' but in the form: 'Shall we prefer a language of physics (and science in general) that contains theoretical terms, or a language without such terms?' From this point of view the question becomes one of preference and practical decision."[6] Although expressed in the language of "Empiricism, Semantics, and Ontology," this application of the distinction between theoretical and practical questions to the realism-instrumentalism controversy is a consequence of Carnap's commitment to the partial interpretation account of theories, rather than a necessary feature of the view, developed in his essay, that insofar as certain traditional philosophical questions about the existence of abstract entities are meaningful, they are practical questions of framework choice.[7]

There is an unacceptable asymmetry between the partial interpretation approach to existential hypotheses involving the unobservables associated with molecular and atomic reality and its approach to theories expressing laws of nature: While acknowledging that theories expressing laws always remain open to revision and subject to qualification, advocates of partial interpretation do not hold that there is anything inherently problematic about the assertion that such theories are true. However, theories which include existence claims about unobservables were typically treated differently: In order that the assertion of such a claim should not be misunderstood and misleadingly interpreted as a claim "about reality," it was thought necessary to qualify it with the proviso that it must be understood in terms of our acceptance of a linguistic framework.[8] The point of this proviso was that assertions about the reality of unobservable entities were held to be the covert expression of a preference for theoretical language, and, in the case of particular interest to us, a preference for using the theoretical language of atoms and molecules to guide our expectations about observable events.[9]

It is difficult to trace the source of this asymmetry, but it seems plausible to suppose that it stems from the perception that the methodology by which hypotheses asserting the existence of unobservable entities are secured is importantly different from the one to which we appeal in securing laws. If we take the classical corpuscularian philosophers as our model, the introduction of unobservable entities proceeds by the method of hypothesis or postulation, and the properties attributed to the postulated entities might well be accessible *only* insofar as they fulfill the purposes that are mandated by explanations which appeal to them. Our examination of Perrin's argument for molecular reality will show that whatever the extent to which this may have been true of early advocates of atomism, it is by no means the only method at our disposal. What Perrin called his "connecting link" between visible particles and molecular reality is justified on the basis of an argument that extrapolates from what is known of Brownian particles of various sizes to what should hold for particles which though invisible, are only a few orders of magnitude smaller. As such, it is no less secure than the inductive extrapolation we make when, having determined the value of a parameter from a variety of sources, we conclude that functional dependencies involving it which we have yet to discover will agree with what we have learned so far. Perrin's connecting link to molecular reality allowed him to determine with confidence many properties of molecules and, in doing so, gave to the assumption of their existence a security that went far beyond anything possible on the basis of merely "hypothetical" reasoning.

The account of theories presented here is continuous with the critical use empiricists have traditionally made of the demand that the concepts with which we seek to represent reality must be empirically well founded. There is a pervasive difficulty with the partial interpretation account's implementation of this demand. That implementation was imbedded in a view of the methodological basis of existential hypotheses that was transcended by the physics community long before the partial interpretation reconstruction was first advanced. The missing element in its account of this methodology is theory-mediated measurement. Because of the conflicting demands that the doctrine of partial interpretation places on the notion of a correspondence rule, the empirical well-foundedness that derives from theory-mediated measurement is only very inadequately captured by this doctrine: Correspondence rules must be modest in what they presuppose regarding theory and experiment in order to contribute to an explanation of the meanings of statements containing terms of which we are supposed to have virtually no prior understanding. But to represent the methodology by which we actually achieve epistemic access to a theoretical domain, it is necessary to appeal to principles that incorporate a great deal of knowledge of theory and experimental design, and, by implication, an understanding of the relevant theoretical vocabulary. How understanding of theoretical vocabulary is achieved is a subtle, difficult, and largely empirical question, but it is one which a methodological investigation of the empirical basis for theoretical claims needs to set to one side in order to progress to an appreciation of the methodology by which we achieve epistemic access to theoretical domains.

It is not necessary to be a committed scientific realist to be dissatisfied with logical empiricism's representation of the controversy between realists and instrumentalists as a practical one. It will emerge from our discussion that one can consistently maintain a strict neutrality about the metaphysical questions of ontology that Carnap isolated in "Empiricism, Semantics, and Ontology," a reasonable skepticism about the truth of theories, and an understanding of the controversy over the ontological status of entities that transcend observation as one that concerns a question about reality with a perfectly determinate and positive resolution within physics. But to do so, it is necessary to appreciate how and why the evidentiary framework within which theories are situated is capable of generating a much more compelling case for the truth of existence claims involving unobservable entities than is possible within the framework favored by advocates of partial interpretation.[10]

By restricting the discussion of realism to antirealist alternatives like "Carnapian deflationism" and deeply skeptical views about existence claims involving entities which transcend observation, it is possible to avoid a number of issues that a more general discussion of scientific realism would otherwise require. First, it allows us to set entirely to one side any commitment to the view that theories are "typically true," or even "approximately true"—without, however, precluding the possibility that progress in science is cumulative and that theories are often retained as limiting cases of the theories which replace them. Second, it divorces our

investigation from the program of providing a general explanation of the "success of science" in terms of the notions of truth
and reference. And third, it makes clear that it is not a part of our
project to advance a "hermeneutical interpretation" of scientists'
pronouncements about "the aim" of science.

What distinguishes the view presented here from various theses
associated with scientific realism is the contention that to address
logical empiricist conceptions of the realism-instrumentalism
debate, together with various skeptical views of existence claims
involving unobservable entities, it is necessary and sufficient to
refine our understanding of the method of argument by which
such claims have, in particularly influential cases, been justified.[11] The two examples on which I focus in order to establish this
contention are Perrin's argument on behalf of the molecular hypothesis and the complementary contributions of J. J. Thomson
involving the corpuscular nature of cathode rays. Modulo the
usual caveats that apply to the strength of the conclusions that may
be drawn from an empirical investigation, I hope to show why
we are justified in supposing that the considerations advanced
by Perrin and Thomson marked a turning point—a turning
point distinguished by Poincaré's conversion to atomism—that
set the path which culminated in the demonstration of the truth
of their respective existential hypotheses. It will follow that Perrin's and Thomson's contributions undermine the thesis that the
controversy between realists and instrumentalists is at best the
expression of a mere preference for or against theoretical vocabulary, and it will also follow that their contributions undermine
various skeptical views of existential hypotheses involving unobservable entities. The present study is an extended argument
against logical empiricism's development of these doctrines and an
articulation of an alternative account of the epistemological basis

of hypotheses regarding the nature and constitution of matter in the form of existence claims involving entities which transcend observation.

The discussion of Perrin's and Thomson's contributions to our understanding of the status of existence claims regarding molecular and subatomic reality addresses the realism-instrumentalism debate as it was framed by the logical empiricists, and as it was somewhat imperfectly addressed by their scientific realist critics. But the current interest in realism, at least in the philosophy of physics, is dominated by whether it is possible to defend a "realist" interpretation of quantum mechanics. The issues this raises are very different from those involving the status of existence claims about molecules and subatomic particles—so different that antirealist participants in this debate can, and, with only a small number of exceptions, *do* concede the atomic constitution of matter.

The question that quantum mechanics raises for realism is whether the theory is susceptible to an "interpretation," according to which the theory is more than a useful formalism for predicting measurement results. The point of an "interpretation" of quantum mechanics is to isolate the theory's conceptually novel contribution to the development of physical theories. An instrumentalist interpretation holds that the conceptual novelty of the theory consists in the fact that quantum mechanics can *only* be understood as a useful formalism for anticipating experimental results. By contrast, a realist interpretation rejects this deflationist view of the theory.

In addition to *interpretations*—which do not change the basic structure of nonrelativistic quantum mechanics, as represented by von Neumann's Hilbert space axiomatization—there are realist *extensions* of the theory which supplement it in order to clarify the theory's compatibility with one or another form of realism. As we will see in Chapter 4, among "extensions," there is a useful distinction to be drawn between realist extensions of the *theory* and realist extensions of the quantum mechanical *state* of a physical system. But for the present, we can ignore this refinement. The acceptability of either modification is rejected by instrumentalist conceptions of the theory's unique contribution to the conceptual structure of physics, while to some—but not all—opponents of instrumentalist interpretations, the theory can be understood realistically only under some such modification.

The orthodox interpretation of the theory, developed in different ways by Bohr, Heisenberg, and others during the heyday of logical empiricism, has always been generally regarded by its critics as explicitly positivistic. Today the orthodox interpretation is widely dismissed by the philosophy of physics community on the ground that it is "merely" an instrumentalist interpretation of the theory. I believe that our discussion of Perrin and Thomson and the methodology of theory-mediated measurement can be exploited to argue that, at least in the case of Bohr's much-maligned thesis regarding the primacy of classical concepts, the situation is far subtler than such a characterization suggests. The conception of the evidentiary framework of physics to which the appreciation of the methodology of theory-mediated measurement will lead us both extends and supports this thesis. The thesis of the primacy of classical concepts can be incorporated into an interpretation of the quantum theory that is continuous with our conception of physical theories as contributions to our representation of the structure and constitution of reality.

The quantum theory does indeed make a unique contribution to our conception of physical theories, but its contribution does not consist in renouncing the basic realist intuition that the

instrumental value of theories of physics goes beyond the prediction of measurement results to include the revelation of salient truths about reality. The conceptual novelty of quantum
mechanics consists in its isolation of a basic structural feature of
our theoretical framework that had not previously been recognized as an object for theoretical reflection and revision. In this
respect, it has a close affinity with special relativity, especially
under the understanding of relativity to which we have become
accustomed since Minkowski. Prior to special relativity, it was
supposed that the universality of the Lorentz group is based on
the electromagnetic constitution of matter: If matter is fundamentally electromagnetic, then the appropriate symmetry group
is given by Maxwell's theory, and the phenomena of time dilation and length contraction are to be understood as dynamical
effects whose source is explained by what Einstein called a "constructive theory,"[12] in the present case, a constructive theory of
the constitution of matter that incorporates Maxwell's laws.

The first step toward the recognition that the structure of
space-time is an object of theoretical reflection was taken when
Einstein reconceptualized the universality of the Lorentz group
by deriving it from a relativity principle and the constancy of the
velocity of light. For Einstein, the Lorentz group is the symmetry
group of all inertial motions *regardless of the physical constitution of what is moving.* Minkowski's contribution was to show
how, from this perspective, the Lorentz group incorporates a conception of the structure of space-time that differs fundamentally
from the conception implicit in pre-relativistic physics. In particular, time dilation and length contraction are features of the
Minkowski structure of space-time and do not depend on any
particular theory of how matter is constituted.[13]

Although the flat Minkowski space-time of special relativity was later replaced in the general theory by a space-time of variable curvature, the significance of the turn to Minkowski geometry is preserved by its retention in general relativity as the structure of space-time "in the small." Hence our general point regarding the instrumental value of theories, in leading us to salient truths about reality, is as true of the principles of the special theory and the geometry of space-time as it is of constructive hypotheses like the molecular hypothesis and the constitution of matter.

Taking the example of special relativity and Minkowski space-time as a model, I develop the idea that the conceptual novelty of the quantum theory consists in what it reveals about the algebraic structure of physical properties and the notion of probability to which this structure directs us. This algebraic structure lies at the center of von Neumann's Hilbert space formulation of the theory. However, its divergence from classical ideas presents a uniquely difficult problem of theoretical interpretation, one that was only partially addressed by Bohr's thesis of the primacy of classical concepts. Our review of the dialectic involving Bohr, Einstein-Podolsky-Rosen, Einstein, and J. S. Bell forms a necessary component of the proposal that the contribution of quantum mechanics to our representation of physical reality is a principle-theoretic one regarding the nature of probability and the algebraic structure of the properties to which it applies. So also do foundational results about Hilbert space that came long after von Neumann's axiomatization. These matters, and the account of the conceptual novelty of the theory that they suggest, conclude our study of theories.

1 Logical Empiricist and Related Reconstructions of Theoretical Knowledge

1.1 The Partial Interpretation Account of Theories

The partial interpretation account of theories holds that the vocabulary of a theory consists of an observational and a theoretical component, where the application of the observation-theory distinction is based on whether a vocabulary item is held to apply to entities in the intended domain of the theory which are observable (in which case the item belongs to the *observational vocabulary*) or unobservable (in which case it belongs to the *theoretical vocabulary*).[1] In its classical formulation, the distinction partitions vocabulary items into just these two classes.[2] (The possibility of a third class of terms—*mixed vocabulary* items that are understood to apply to both observable and unobservable entities—will be addressed later in this chapter .) Given this distinction in vocabulary, the sentences of the language of a theory are divided into three classes: one consisting of sentences which are generated from just the observation vocabulary, another of sentences generated from just the theoretical vocabulary, and a third consisting of sentences generated from the combined observation and theoretical vocabularies. These are, respectively, the *observation sentences, theoretical sentences,* and *correspondence rules* of the language. A *theory* is the conjunction of a selection of theoretical sentences and correspondence rules; I will ignore various possible generalizations of the notion of a theory that are irrelevant to the conceptual issues on which I will focus. There are no special restrictions on the logic of a theory, and it may be either first order or higher order.

Historically, the partial interpretation account of theoretical knowledge derives from the idea that theoretical terms are introduced by sentences which, taken by themselves, are indistinguishable in their epistemic status from the statements of a pure mathematical theory. Of chief importance to the partial interpretation account is the notion that theoretical statements share with the statements of a mathematical theory the property that their interpretation is responsible only to the logical category of their constituent nonlogical constants. This view of theoretical statements is a consequence of the fact that the partial interpretation account is a continuation and extension to the theories of physics of the axiomatic tradition that Hilbert initiated in pure mathematics. Especially influential was Hilbert's contention that the primitives of a mathematical theory are whatever satisfies its axioms. This contention—that the postulates of a theory "implicitly define" its primitive notions—swept away the subjective associations that characterized an older tradition's understanding of a mathematical theory's primitives, even in the case of geometry, where they were thought to have a familiar "intuitive" content.[3] The partial interpretation account sought to extend Hilbert's analysis of mathematical theories to physics by providing an account of the empirical content of the theoretical statements of physics based on the connections between theoretical terms and observation terms that are expressed by correspondence rules.

One can see in this brief sketch the two characteristic theses of the partial interpretation view: the first, its claim that only the observation vocabulary is completely understood; and the second, the correlative claim that the interpretation of the theoretical vocabulary is limited by constraints which depend only on the logical category of the theoretical terms and whatever restrictions the true observation sentences impose on the domain of unobservable entities over which the theoretical sentences and correspondence rules are evaluated. I will refer to this second claim as the *structuralist thesis*. We have yet to explain how the partial interpretation view conceives the relation between interpretations, true interpretations, and truth.

1.2   Carnap on Ramsey Sentences and the Explicit Definition of Theoretical Terms

Carnap's mature reconstruction of the language of science[4] builds on and extends the partial interpretation view of theories. The central notion of this account is the *Ramsey sentence* of a theory: the sentence formed by replacing theoretical terms by (new) variables of the appropriate logical category, then closing the resulting formula by adding an existential quantifier for each of the new variables. It is a very short step from the two characteristic theses of the partial interpretation account of theories to the notion that a partially interpreted theory's Ramsey sentence captures its "factual content": the Ramsey sentence is observationally equivalent to the theory in the sense that any argument from the partially interpreted theory to a sentence of the observation language can be recovered using the Ramsey sentence instead; and the Ramsey sentence's use of variables in place of uninterpreted theoretical terms simply makes explicit the commitment of the partial interpretation account to the structuralist thesis. As Ramsey expressed it:

> So far . . . as *reasoning* is concerned, that the [transforms of the theoretical sentences and correspondence rules in the matrix[5] of the Ramsey sentence of the theory] are not complete propositions makes no difference, provided we interpret all logical combinations as taking place within the scope of a [single existential] prefix. . . . For we can reason about the characters in a story just as well as if they were really identified, provided we don't take part of what we say as about one story, part about another.[6]

Carnap's mature reconstruction refines the doctrine of partial interpretation in two principal respects. As we have already noted, Carnap explicates the factual content of a partially interpreted theory in terms of its Ramsey sentence. But Carnap took things a step further by combining his account of the factual content of a theory with an explication of theoretical analyticity—analyticity relative to a theory—in terms of what has come to be known as the *Carnap sentence* of a theory: the conditional whose antecedent is the theory's Ramsey sentence and whose consequent is the partially interpreted theory. Before Carnap, the distinction between the factual and analytic (and hence, nonfactual) components of a theory followed the distinction between postulates and definitions. But since this distinction is inherently arbitrary, its utility for a dichotomy that is supposed to reveal our factual commitments may be doubted.

The Carnap sentence is justifiably regarded as analytic because it is a kind of "implicit definition" of the theoretical vocabulary, one that is provably nonfactual in the sense that the only observation sentences it logically implies are logical truths. And as John Winnie (1970) later showed, the Carnap sentence, like a proper definition, satisfies a special noncreativity condition similar to the noncreativity condition that is customary for proper *explicit* definitions.[7]

Carnap advanced the Ramsey sentence not just as a clarification of the partial interpretation view of theories but also as a correct representation of how scientists understand their theoretical claims. They intend, Carnap held, an "indeterminate" claim, one that may have many interpretations under which it comes out true. As scientists understand them, theoretical claims are indeterminate as to the interpretation of their theoretical vocabulary, and any representative class or relation which makes true the Ramsey sentence of the theory to which the claim belongs is as acceptable as any other. To narrow down the interpretation any further than is demanded by the truth of the Ramsey sentence would, for Carnap, violate the intentions of the scientist who constructed the theoretical system.

In one of his last papers on the subject,[8] Carnap converts the *implicit* definition of theoretical terms by the Carnap sentence into a sequence of *explicit* definitions of them. But these explicit definitions do not eliminate—and were not intended by Carnap to eliminate—the indeterminateness of his earlier account. Indeed, Carnap formulates his explicit definitions in what he calls a "logically indeterminate" language. The language L_ε which he employs is a standard first- or higher-order language enriched with Hilbert's epsilon operator and the extensional axioms which govern its use. There are just two such axioms. Given a formula *Fx* in one free variable, the first axiom tells us that if there is something satisfying *Fx,* then there is an "ε-representative" of *F,* denoted "$\varepsilon_x(Fx)$," that is selected by the choice function which interprets the epsilon operator. The second axiom tells us that if the formulas *Fx* and *Gx* are extensionally equivalent, their ε-representatives are the same.

That it should be possible to apply Hilbert's epsilon operator to the Ramsey sentence reconstruction of theories is a consequence of Carnap's observation that the Carnap sentence of a theory can be derived from a sentence that is in the same form as the first of the axioms governing the epsilon operator. This sentence can be understood as asserting that if there is a sequence of *classes* of the appropriate type which satisfies the matrix of the Ramsey sentence of a theory, then there is an *ε-representative* such sequence (i.e., a sequence consisting of *n ε-representative classes,* where *n* is the number of new variables that were introduced when the partially interpreted theory was replaced by its Ramsey sentence). Carnap observed that if the theoretical terms $_{Ti}(1 \leq i \leq n)$ are now explicitly defined as the ε-representatives of such a sequence, the Carnap sentence follows.

For Carnap, the principal virtue of his proposal is that it incorporates the convenience of having the use of a theoretical vocabulary while retaining all the characteristic indeterminateness of that vocabulary, which is the hallmark of the partial interpretation view and of his mature reconstruction in terms of Ramsey and Carnap sentences. Carnap writes that the theoretical postulates and correspondence rules "are intended *by the scientist who constructs the system* to specify the meaning of [a theoretical term] to just this extent: if there is an entity satisfying the postulates, then [the term] is to be understood as denoting one such entity. Therefore the definition [of a theoretical term by means of the epsilon operator] gives to the indeterminate [theoretical term] just the intended meaning with just the intended degree of indeterminacy" (Carnap 1961, p. 163; emphasis added).

1.3   A Proposal of David Lewis and Two Theorems of John Winnie

David Lewis's (1970) article is sometimes credited with having refined Carnap's and Ramsey's reconstructions and to have improved on Carnap's approach to the explicit definition of theoretical terms by showing how it might be possible to avoid multiple interpretations of the theoretical vocabulary under which the theory comes out true. Lewis maintained that allowing for what he calls "multiple realizations" concedes too much to instrumentalism. Lewis does not say why multiple realizability is a concession to instrumentalism, but let us for the moment grant the point and consider how he makes the case that there are many theories which, if realizable, are *uniquely* realized. Lewis is clear that he must provide an independent defense of this contention, since the possibility of there being just one realization appears to be excluded by two theorems of John Winnie (1967). Modulo the conceptually unimportant technical restriction that not all theoretical properties and not all theoretical relations are universal, Winnie shows that on the partial interpretation view of theories, if a theory has one realization, there is always another; and if a theory is realizable at all, it is arithmetically realizable.

Lewis's response to Winnie rests on two features of his conception of the language in which theories are formulated. First of all, Lewis follows the partial interpretation account by dividing the vocabulary of a theory into two parts, which he calls the "O-vocabulary" and the "T-vocabulary" of the theory. However Lewis's "O-T distinction" is not the distinction between observational and theoretical vocabulary of the partial interpretation account. Lewis's distinction concerns *old* vocabulary, vocabulary which is understood prior to the formulation of the vocabulary-introducing theory; and the contrast Lewis's distinction draws between old and T- or *new* vocabulary has nothing to do with observation or observability. In principle, Lewis's O-T distinction could be completely orthogonal to the observational-theoretical distinction of Carnap and the doctrine of partial interpretation. A second, related, difference involves Lewis's notion of an "O-mixed term." This is a notion that does real work for Lewis, but before explaining it, some further background regarding the partial interpretation view and its relation to Lewis's O-T distinction is necessary.

In his exposition of the partial interpretation reconstruction, Winnie includes in addition to the observational and theoretical predicates a separate category of *mixed predicates,* predicates that apply to both observable and unobservable entities. Lewis has the notion of a *mixed term,* and of special importance are those he calls "O-mixed" terms. These are terms which, like Winnie's mixed predicates, can apply to both observable and unobservable entities. But their characterization—unlike Winnie's characterization of observational and theoretical predicates—has nothing to do with observability. And while Winnie's mixed predicates are distinguished from observation predicates, Lewis's O-mixed terms count as O-terms, and as such, are assumed to be fully understood whether they apply to observable or unobservable entities; therefore the interpretation of O-terms—whether they are unmixed and refer only to observable entities, or are mixed and refer also to unobservable entities—must be preserved as we pass from one realization of a theory to another.

The situation is altogether different for Winnie and for the standard view. When Winnie defines the permutation map which establishes the existence of alternative realizations, it is only the entities in the observable part of the domain that cannot be permuted, and it is only the interpretation of the observation predicates—predicates which apply only to observable entities— that must be the same in any intended realization of a partially interpreted theory. No such requirement applies to the interpretation of theoretical predicates; nor does it apply to mixed predicates.

Since Winnie's permutation map is the identity on observable entities, it is trivially true that observable relations are isomorphic to their images under his mapping. But although it is trivially true that Winnie's map is an isomorphism between observable relations, it is not part of Winnie's argument that *every* one-one map from the observable part of the domain onto itself can be extended to an isomorphism on the properties and relations which interpret the observation predicates of the theory. However, the situation is different for one-one maps from the unobservable part of the domain onto itself and the relations which interpret the theory's theoretical predicates.

To establish that a partially interpreted theory has many models if it has one model, Winnie requires a map that permutes at least one pair of unobservable entities in such a way that it changes the image, under the mapping, of the interpretation of at least one theoretical predicate. But on the partial interpretation view—and this is the observation Winnie's proof rests upon—the theoretical predicates and relations can always be understood so that the relations which interpret them *must* be isomorphic to their images under an *arbitrary* one-one mapping from and onto the unobservable part of the domain. As for the mixed properties and relations, they are constrained by the condition that any mapping which defines them cannot be arbitrary on the observable part of the domain. Thus, if a partially interpreted theory has a model M, it has another model M*, where M* is formed by *defining* its theoretical and mixed properties and relations as the images of the theoretical and mixed properties and relations of M under a permutation map which is the union of the identity map on observable entities and any suitable (i.e., suitable for the purposes of his theorem) one-one mapping from and onto the unobservable entities. Since the theoretical and mixed properties and relations of M* are *defined* as the images of the theoretical and mixed properties and relations of M under this permutation map, it follows that M and M* are *by construction* distinct models with isomorphic properties and relations.[9] The admissibility of Winnie's interpretation of the theoretical predicates in M* is evidently a consequence of the second characteristic feature of the partial interpretation view, namely, the fact that according to the structuralist thesis, the interpretation of theoretical predicates over the unobservable part of the domain is restricted by constraints which depend only on their logical category. Since the properties and relations of M are isomorphic to their images in M* under the permutation map, M* is a new model of the partially interpreted theory if M is a model of the theory.

To see why this argument does not affect Lewis's claims about uniqueness of realization, let us suppose that electrons and protons are unobservable entities, that electrons are (strictly) smaller than protons, and that "smaller than" is a term of Lewis's O-mixed vocabulary; any realization must therefore preserve the interpretation of "smaller than." It follows that the construction of a realization which, like Winnie's, interchanges an electron and a proton, is ruled out. In fact, the only case in which Lewis's and Winnie's notions of realization coincide is the case in which Lewis's O-terms and T-terms have (respectively) only observable and only unobservable entities in their extensions, and where therefore there are no mixed terms. In such a case, even on Lewis's account, a theory must have more than one realization if it has any realization at all. Hence, Lewis's approach to theories and the definition of theoretical terms is at best only "accidentally" affected by Winnie's results.

To sum up our discussion of Winnie and Lewis, Winnie's mixed predicates form a special category distinct from observation predicates, and only the observable entities in the interpretation of mixed and observation predicates are unaffected by his permutation map. But for Lewis, mixed terms can be classified as O-mixed terms, and in order to avoid Winnie's permutation argument, it suffices that there should be a suitably rich collection of O-mixed terms. In fact it may suffice that there should be *one* term that stands for a relation such that all the various types of theoretical entity are comparable in terms of this relation. In our example of electrons and protons, it is sufficient that the particles of one kind are (strictly) smaller along some dimension than those of the other—assuming, of course, that *smaller than* along this dimension is picked out by an O-mixed term. It is therefore not at all implausible that for Lewis, a realizable theory can always be extended by the introduction of O-mixed terms and appropriate postulates involving them to a theory that is *uniquely* realizable.

Lewis's deployment of the Ramsey sentence and a modified form of the Carnap sentence shares a strong formal affinity with

Carnap's reconstructions: Lewis's adoption of the Ramsey sentence suggests a commitment to the idea that a model of our understanding of new or T-terms—including the case in which the new terms are associated with the introduction of a new class of entities—is adequate if it captures the inferential connections of the statements containing the new terms with the sentences in the O-vocabulary. But although Lewis's old vocabulary may well include "observation" terms, the connections between T-terms and O-terms which Lewis's mixed statements express is not explicitly proposed as a connection with observation vocabulary, but merely with vocabulary that is understood. And for Lewis, by contrast with the partial interpretation or Ramsey-sentence reconstructions of the logical empiricists, so far as our understanding of *old* vocabulary is concerned, it might be based entirely on our grasp of the inferential connections into which its items enter—independently of whether or not these include inferential connections with observation sentences.[10] It is clear therefore that despite their formal affinity, the absence of an epistemological motivation underlying Lewis's reconstruction of theories makes his account very different from both the partial interpretation reconstruction and from Carnap's various refinements of it.

Carnap and the advocates of partial interpretation take it as a desideratum of an adequate reconstruction of theoretical knowledge that it should address the empirical basis of theoretical claims. On the partial interpretation reconstruction, this problem is addressed by the provision of an explanation of our understanding of theoretical claims in terms of the connection correspondence rules establish between theoretical and observation vocabulary. Carnap's and Ramsey's Ramsey-sentence reconstructions dissolve the problem of how we come to understand the meanings of terms which apply to unobservable entities by *eliminating* theoretical terms in favor of variables. But this dissolution of the problem is merely an emendation—not a rejection—of the partial interpretation view, an emendation that preserves the structuralist thesis. Indeed, Ramsey's and Carnap's use of the Ramsey sentence is based on their recognition that the partial interpretation view subscribes to this thesis. For if, in addition to whatever restrictions the true observation sentences impose on the domain, the constraints on the interpretation of theoretical vocabulary appeal only to the logical category of the theoretical terms, then there can be no objection to their replacement by variables, and the problem of accounting for how theoretical terms are understood then simply disappears. As for the empirical basis for theoretical *claims*—as opposed to our understanding of theoretical *vocabulary*—and the explanation of their difference from the claims of pure mathematics, modulo the elimination of theoretical vocabulary, these issues are addressed by Carnap's and Ramsey's Ramsey-sentence reconstructions much as they are by the partial interpretation reconstruction. Instead of appealing to the association of theoretical claims with observation sentences by the mediation of correspondence rules, the empirical basis of such claims is accounted for by the association of the Ramsey-sentence transforms of theoretical claims with observation sentences that is effected by the Ramsey-sentence transforms of correspondence rules.

The centrality of the problem of the empirical basis of theoretical claims to the logical empiricists' reconstruction of theories serves not only to distinguish their approach from Lewis's. It also shows why it is so misleading to characterize their account—whether partial interpretation, Ramsey sentence, or one of Carnap's later reconstructions—to have incorporated a "syntactic view" of theories.[11] For the logical empiricists, theories are linguistic objects. But to call the logical empiricist account "syntactic" misses the fact that it is first and foremost a reconstruction of theoretical knowledge that purports to show how observation bears on the empirical character and evidential support of theoretical claims. Evidently, neither goal can be successfully addressed without going beyond syntax—as indeed the principal logical empiricist reconstructions do by assuming that the observation language is interpreted.

This contrasts with the situation in mathematical logic, where a theory is defined as a set of sentences in a language about which we assume *only* that its syntax and underlying logic are completely explicit. The notion of a theory which arises in mathematical logic—what I will refer to as *the logical tradition's notion of a theory*—is also motivated by an epistemological problem which is no less fundamental to its notion of a theory than the epistemological motivation underlying logical empiricism is to *its* conception of a theory. The logical tradition sought to show that a *finitary* notion of proof suffices for the reconstruction of all mathematical reasoning—even reasoning within theories whose intended interpretation is over an infinite domain of arbitrarily large cardinality. But here the restriction to syntax in the characterization of a theory is entirely natural, since it is essential to the successful positive solution of the epistemological problem which motivates the logical tradition that a theory should be represented as a purely syntactic object. Although the logical empiricist approach to theories was profoundly influenced by the logical tradition, its goals were different: It sought to build a platform for the representation of the theoretical claims of physics that would be capable of illuminating their content and the basis on which they are understood and evaluated. In particular, it sought to show how observation must be a central component of an adequate empiricist solution to this problem. The questions that the logical empiricist approach sought to address demanded—and were recognized by its proponents as demanding—a notion of theory that includes more than the purely syntactic conception of the logical tradition.

1.4 Putnam's Model-Theoretic Argument

We have seen how, by contrast with Lewis, Carnap was prepared to accept multiple realizability as a point *in favor* of partial interpretation and Ramsey-sentence reconstructions. In his "Replies" in the Schilpp volume devoted to his work Carnap even went so far as to endorse an arithmetical interpretation of the Ramsey sentence of a theory as the *correct* understanding of what, on his reconstruction, the sentence asserts.[12] In doing so, he might be understood to have also anticipated and embraced the content of the second of Winnie's two theorems as an acceptable consequence of his view of the factual content of theories. In light of these considerations, let us for the moment set to one side the issues connected with arithmetical interpretations and multiple realizability and turn our attention to a closer examination of the question: How do theories, whose characteristic feature is that their theoretical claims transcend observation, acquire their empirical status? This brings us to Hilary Putnam's model-theoretic argument, by which I mean his first such argument, the one which purports to show the incoherence of the idea that an "epistemologically ideal" theory might be false. This argument was first presented in Putnam's 1976 American Philosophical Association presidential address. However I should emphasize that my interest is restricted to the *actual argument*; I will not address any of Putnam's uses of his argument, which are all more various than the application I will isolate. As I understand its significance, the argument shows that the answer to our question given by the doctrine of partial interpretation and its close descendants is incompatible with the thesis that when theories which transcend observation are true, they express salient truths about unobservable entities. The fault with all these views stems from their failure to satisfactorily address the basis for our epistemic access to theoretical domains.

The model-theoretic argument consists of a simple technical argument and an observation. The technical argument establishes that any model of a theory's observational consequences can be extended to a model of the theory's theoretical sentences and correspondence rules, where the domain of this extension is the standard domain of observable and unobservable entities. The argument which establishes this conclusion also supports an observation, namely, that on the partial interpretation reconstruction of theories, the conditions under which a partially interpreted theory can be shown to be satisfiable suffice to show that the theory is *true*.

Let M be a model of the observational consequences of our theory such that the domain of M is the "standard domain" of observable entities. To obtain a model of the observation sentences, theoretical sentences, and correspondence rules, we can exploit a folklore result reported by van Benthem that assures us that there is an "abstract" model N which is an extension of M to a model of the whole theory.[13] Given N and a single nonlogical assumption, it is possible to define a model M* of the theory which, like N, is also an extension of M but has as its domain the standard domain of observable and unobservable (i.e., theoretical) entities. On the "purely contingent" assumption that the domain of M* is of the same cardinality as the domain of N, there is a one-one onto mapping which is the identity on the common observable part of the domain of M* and the domain of N, and is arbitrary from the "unobservable" part of N onto the unobservable part of M*. Let the observable relations of M* be the same as those of N, and define theoretical relations for M* as the images, under this one-one correspondence, of the theoretical properties and relations which interpret the theoretical predicates of the theory in N. This defines a model M* which extends M by the addition of theoretical relations which are defined over the standard domain of unobservable entities, and which are *by definition* isomorphic to the theoretical relations of N; since M* extends M if N does, and since N is a model of the theory, so also therefore is M*. But since this construction of the theoretical relations of M* meets all of the conditions that the partial interpretation account is capable of imposing on such relations, the fact that M* models the theory suffices to show that the theory is not merely true in M*, but *true*.

The arguments of Winnie and Putnam both exploit the same technical idea in their respective definitions of the theoretical relations which interpret the theoretical predicates of a partially interpreted theory. But their arguments are most naturally understood to support conceptually distinct difficulties for the view. My use of Putnam's argument is not directed at the existence of *multiple* realizations; nor does it concern the existence of an *arithmetical* model of a partially interpreted theory. My claim is rather that Putnam's argument takes us from a cardinality assumption, and the existence of what might well be an arithmetical model of the kind explored by Winnie, to the conclusion that, on the partial interpretation view, the fact that a theory is *satisfiable* over the standard domain of observable and theoretical entities suffices to show that it is *true*.

The significance of the model-theoretic argument has been the subject of an extensive discussion. But whatever the resolution of the many controversies the argument has generated involving "metaphysical realism," "intended" reference, or the "indeterminacy" of reference, it seems clear that the notion that the truth of what is asserted about unobservable entities might depend only on their *number* runs counter to one of the simplest and least contentious convictions of "realism" and, indeed, of common sense. This is the conviction that if a theory is true, this is because its theoretical claims have captured a salient aspect of the reality they seek to describe, an aspect that goes beyond any mere question of cardinality.

The partial interpretation account of theories claims to reconstruct the empirical status of a theory's theoretical statements using only the theory's logico-mathematical framework and the apparatus of correspondence rules. But the fact that when we are restricted to just these resources the truth of theoretical claims reduces to their satisfiability in any sufficiently large model of the true observation sentences, shows that the reconstruction has failed to correctly represent the nature of the epistemological status of a theory's theoretical claims. It has failed because the epistemic basis for such an assertion of satisfiability is entirely different from what is required by an assertion of truth. The idea that the claim that a theory is true should depend only on a cardinality constraint, and a logical argument fails to adequately separate the epistemic basis for the truth of the theoretical assertions of an empirical theory from the epistemic basis for the mere satisfiability of the "abstract" assertions of a purely mathematical theory over a given domain.

The conclusion we have just reached should perhaps have been anticipated, given the origin of the partial interpretation view in Hilbert's conception of the foundations of geometry. In his correspondence with Frege, Hilbert defended the idea that satisfiability in a sufficiently large domain is a suitable surrogate for the "truth" of a mathematical theory. But whatever its plausibility for theories of pure mathematics, the methodological demands we impose on the theoretical claims of physics cannot be captured by so weak a requirement, at least not if we wish to preserve the methodological difference between physics and pure mathematics. An advocate of partial interpretation might respond to this objection by recalling that a physical theory will qualify as true not if it is merely satisfiable, but only if it is satisfiable in a model which is an extension of a domain that forms the basis for a model of the true observation statements. By contrast, the domains which bear witness to the "truth" of a mathematical theory need not have any connection with such a model. For an advocate of partial interpretation, the theories of physics are true because they are *empirically adequate* in the sense that they have observational consequences, all of which are true; but a theory of pure mathematics is not necessarily associated with *any* observation language and is not required to be empirically adequate.

However, this response misses the point of the model-theoretic argument as we have presented it: Provided the domain over which a partially interpreted theory involving unobservable entities is interpreted includes the domain of the model of the true observation sentences, it is a consequence of the partial interpretation view that the method of argument by which we are able to establish the "truth" of a purely mathematical claim over a given domain also suffices to establish the truth of a theoretical claim.

The model-theoretic argument puts us in a position to see why—pace Lewis—the multiple realizability which afflicts partial interpretation and Ramsey-sentence reconstructions is largely tangential to the question of realism. For suppose that we are given a realization of the sort that Putnam's argument shows is possible. Given such a model, we have seen how, following Lewis (1970), we can rule out alternative realizations by supplementing the partially interpreted theory with a judicious selection of O-mixed terms and appropriate assumptions involving them. Notice, however, that this is compatible with the possibility that a theory is true only because it has a realization that models its observational consequences and is the right size. So in light of Putnam's argument, uniqueness of realization is insufficient to ensure the widely held conviction that if our physical theories are true, this is because they succeed in isolating salient truths about the entities with which they deal, independently of whether these entities are observable or unobservable. In any case, when, long after his 1970 paper on theoretical terms, Lewis came to address Putnam's argument, he did not appeal to O-mixed terms to resolve the problem he took the argument to pose but instead based his reply on a distinction among possible realizations.[14] The core assumption of Lewis's response is that a theory is *true* only if it is true relative to a realization whose properties and relations are *natural*. Since there is nothing in Putnam's construction of his interpretation of theoretical predicates which requires that they should be natural properties and relations, Lewis argued that the construction fails to show that the theory's theoretical claims are, in the relevant sense, true.

It might be thought that we should adapt Lewis's reply to the model-theoretic argument and supplement the partial interpretation account by restricting the class of admissible realizations to those that involve natural relations, thereby distinguishing, in the way Lewis proposes, true theories from theories that are merely satisfiable in a domain that extends a model of the true observation statements. Lewis's distinction could be further exploited to characterize true *empirical* theories as those that are not merely satisfiable in some realization or other, but are true because they are true in a realization whose relations are natural. But we should be cautious about accepting Lewis's suggestion as an adequate response to the model-theoretic argument or as a guide for emending the partial interpretation view.

To begin with, Lewis's reply to Putnam leaves unresolved the problem of how we are able to make significant claims about relations that are *not* "natural." Even if we have no interest in theorizing about such relations, an adequate response to the model-theoretic argument should nevertheless explain how it is *possible* to do so without the assertion of the truth of such a theory collapsing into an assertion of its satisfiability over a domain—even a domain that extends the model of the observational consequences of the "theory" of such a natural relation. Indeed, as Fraser MacBride has remarked, on the assumption that we achieve knowledge of natural relations only with the progress of science—and perhaps only after many distractions involving nonnatural relations—anyone following Lewis's suggestion *must* have an interest in how we manage to make significant, but as it happens, misguided claims about nonnatural relations.

But secondly, and more importantly, addressing Putnam's argument by appealing to Lewis's proposal obscures the difficulty that the model-theoretic argument raises for partial interpretation and Ramsey-sentence reconstructions. The problem with these approaches is not their failure to designate certain properties and relations as natural, but the fact that they are too weak to explain the difference between the epistemological status of theoretical and purely mathematical claims. But then the partial interpretation framework for addressing how theories are warranted must also fail to capture the methodology by which claims about unobservables are established, and this shows that a different approach to these two issues is required.

An emendation of the view based on Lewis's reply to Putnam only succeeds in recording the fact that we *do* distinguish mathematical claims from the theoretical claims of physics, but it has nothing to contribute to our understanding of the methodology by which we make this distinction. Nor does it contribute to our understanding of how we successfully gain epistemic access to theoretical domains in order to warrant our claims about them.

1.5 Ramsey on Russell's Analysis of Matter and the Partial Interpretation of Theories

As Michael Friedman and I have noted, Putnam's argument is reminiscent of an objection raised by M. H. A. Newman against the causal theory of perception Russell advanced in *The Analysis of Matter*.[15] Newman understood Russell to have argued that our knowledge of events with which we are not perceptually acquainted is "purely structural" in the sense that in general, we can know of them only that they are ordered by relations that are structurally similar to the relations that order our percepts—the events with which we are acquainted. Newman observed that a claim of structural similarity is significant when the relations between which it is asserted to hold are specified, but is empty when it consists of the claim that there is merely some relation over a domain to which a given relation is structurally similar: Modulo an assumption about cardinality, it is true as a matter of logic that for any class, there is always *some* relation with an assigned structure that holds between the members of the relevant class. Newman concluded first, that Russell's theory fails to capture the conviction that our assertions about the structure of a class of unperceived events, if true, are not true merely on the basis of logic and an assumption about their number; and second, that the fault lies with Russell's theory of our knowledge of the unperceived part of the world and the notion that our epistemic access to it is restricted to its purely structural properties.

In response, Russell claimed that his considered view was different from the one Newman criticized. He expressed this in a letter to Newman, writing that he had always implicitly assumed acquaintance with spatiotemporal relations that hold among unperceived events, and that his statements to the contrary were unfortunate slips.[16] However, Russell's *AoM* is so replete with explicit endorsements of the thesis that, of the relations that order the unperceived parts of the world, we know only their mathematical properties, that even if this is not the view he was advocating, his remarks to this effect are sufficiently numerous to constitute a reasonably complete formulation and articulation of it.[17]

On the interpretation of the model-theoretic argument that Friedman and I proposed, Putnam's argument is a natural extension of Newman's criticism of Russell's structuralism to partial interpretation and Ramsey-sentence reconstructions of theories involving entities that transcend observation.[18] But more is true: These reconstructions are not only systematically related to what is standardly, if inaccurately, referred to as "Russell's structuralism"; there is also a clear historical link which connects them.

It is a little-known fact that a preliminary formulation of the Ramsey-sentence reconstruction of theories arose in the course of Ramsey's reflections on *AoM*. These reflections, recorded in the fragment, "Physics Says,"[19] were occasioned by Ramsey's reading of the book's first chapter. This is the chapter where Russell discusses the notion of "interpretation" relevant to his study. The fragment is interesting both for what it reveals about Ramsey's understanding of Russell and for the light it casts on the development of his own views. It is clear from internal evidence that the fragment was written at roughly the same time as his posthumously published "Theories," but I have no archival evidence as to its precise date of composition; nor have I been able to ascertain whether Ramsey was familiar with Newman's article or in contact with him when the fragment was written.

By contrast with Newman, Ramsey clearly understood Russell to have allowed that we are acquainted with at least some of the spatiotemporal relations that hold among unperceived events, thereby confirming the central contention of Russell's response to Newman. Ramsey noticed that for Russell *compresence* and *time interval* are examples of relations with which we are acquainted and which hold among both perceived and unperceived events. But he immediately adds that these are insufficient for carrying out Russell's "program of logical construction," and that Russell himself has acknowledged that they are not enough.[20] That program would require for its completion the provision of logical constructions out of Russellian "events" and the various relations among them of all the primitive notions of physics, after the manner of the logical construction of the cardinal numbers in *Principia*. *AoM* sketches how such a construction might go only for the case of space-time points.[21]

Although Ramsey is skeptical about the prospects of providing such an interpretation, his doubts about the likely success of Russell's program of logical construction are only part of the ground—and by no means the most important part—on which he bases his rejection of Russell's notion of interpretation. His central argument against Russell is that an interpretation in Russell's sense can be shown to be unnecessary for explicating either the *empirical* character of physics, or the meaning of the attribution of *truth* to its theories. Ramsey argues that an altogether different and much simpler solution is available, one that dispenses with Russell's logical constructions in favor of Norman R. Campbell's notion of a dictionary relating observational and theoretical vocabulary—the respective vocabularies of what in "Theories," Ramsey calls the "primary" and "secondary" systems.[22]

Campbell's notion of a dictionary is sometimes credited with being a precursor to the correspondence rules of the partial interpretation account of theories. In any case, Ramsey certainly exploits Campbell's idea to justify his rejection of Russell's claim that unless we can find a system of logical constructions of the primitive notions of physics "which gives a due place to perceptions . . . we have no right to appeal to empirical evidence" in evaluating the truth of its theories.[23] Ramsey's point is, of course, *not* that we can ignore the contribution of perception to the empirical status and truth of physics, but that in order to accommodate perception, it is not necessary to provide the logical constructions Russell regards as essential.

Ramsey's argument for this conclusion is very brief. He begins (Galavotti 1992, p. 251), by representing Russell as holding "Physics says = is true if

$$(\exists \alpha, \beta, \ldots R, S): F(\alpha, \beta, \ldots R, S) \tag{1},"$$

where, following the notational conventions of *Principia*, Ramsey uses Greek letters α, β, . . . for class variables and Latin letters R, S, . . . for variables whose values are relations in extension. Ramsey cites *AoM* (p. 8) as the basis for attributing to Russell this schematic representation of physics. And indeed, what Russell writes, when formulating his problem of interpretation—"Given physics as a deductive system, derived from certain hypotheses as to undefined terms, do there exist particulars or logical structures composed of particulars, which satisfy these hypotheses?"—is certainly in keeping with the abstract and schematic representation of a physical theory Ramsey attributes to him. It is also in keeping with the idea that for Russell, the truth of physics consists in there being an assignment of appropriate objects— "particulars or logical structures composed of particulars"—to these variables.

Ramsey next remarks that "the members of α, β, etc. must not be numbers," which I understand to be an allusion to Russell's distinction, earlier in the chapter, between interpretations of different "importance":

> It frequently happens that we have a deductive mathematical system, starting from hypotheses concerning undefined objects, and that we have reason to believe that there are objects fulfilling these hypotheses, although, initially, we are unable to point out any such objects with certainty. Usually, in such cases, although many different sets of objects are abstractly available as fulfilling the hypotheses, there is one such set which is much more important than the others. (*AoM,* pp. 4–5)

To illustrate what is meant by an "important interpretation" Russell turns to the case of geometry, where he argues that although Euclidean and non-Euclidean geometry are interpretable over n-tuples of real numbers, their important interpretation is one which understands geometry to be a part of physics, and, therefore, true of the points of space or space-time, rather than a development of the theory of the real numbers.

Ramsey then observes that since, for a logicist like Russell, *numbers* are logical structures defined in terms of particulars,

Russell's formulation of the problem of interpretation allows for the possibility that the members of α, β, etc. are numbers. Ramsey recognizes that this would not be in accordance Russell's intentions in the case of physics any more than in the case of geometry:

> But the members of α, β, etc. must not be numbers.
>
> Russell says "defined in terms of particulars", but numbers are.

And so, he offers the correction:

> Say rather
>
> $$\alpha = \{x\colon \varphi\,(x)\} \ldots$$
> $$R = \{<x,y>\colon \varphi\,(x,y)\}$$
>
> where the φ's are nonformal [and where perhaps also] F must not be tautological as it is on Eddington's view.[24]

I can only conjecture what more Ramsey may have meant by these very brief remarks. A recurrent theme in his discussion of *Principia* is that while it captures the *generality* of mathematics, it fails to capture its *necessity*.[25] The representation of physics by a general statement like (1) poses the opposite problem of showing why, when the theoretical claims of physics are represented with complete generality, they are not thereby represented as necessary. The requirements of nonformality and nontautologousness are intended to address this issue. To illustrate the point, suppose $F(\alpha, R)$ formalizes the theory of the natural numbers under the relation *less than*. It is logically necessary that the numbers under *less than* form an omega-sequence, but it is a contingent fact whether, under an appropriate binary spatial relation, a sequence of physical objects forms an omega-sequence. Hence, as I propose understanding him to say that if $F(\alpha, R)$ is interpreted as the theory of the natural numbers under *less than,* the propositional functions which determine α and R are what Ramsey calls "formal," and the theory $F(\alpha, R)$ so interpreted is "tautological." But in the case where the theory is interpreted so that it concerns the order type of a sequence of physical objects which are ordered by a particular spatial relation, the functions are nonformal, and the theory $F(\alpha, R)$ of the arrangement of the objects in space is not tautological.

Immediately after offering this correction to Russell's characterization of an interpretation, Ramsey's discussion shifts, and he asserts—without argument—that it is unlikely that Russell can produce a complete system of logical constructions of the primitives of physics because there are too few relations with which we are acquainted to effect the required constructions. Ramsey then turns from the exposition of Russell to a sketch of his own positive proposal and an indication of the nature of its divergence from Russell's view:

Say perhaps "partial interpretation" also perhaps some restrictions on the interpretation of the other variables, i.e. all we know about α, S is not that they satisfy (1).

Any evidence must give us not an interpretation exactly but a dictionary; it must be

Physics I perceive p
I perceive p
Therefore Physics

In exact contradiction to Russell ps. [8] ff.[26]

Ramsey is here in effect endorsing a view of theories along the lines of the partial interpretation reconstruction. To see how and why this is asserted by Ramsey to be in exact contradiction to Russell, recall that in *AoM* (pp. 8–9) Russell distinguishes between a narrower and a wider notion of truth for physics:

> In the narrowest sense, we may say that physics is 'true' if we have the perceptions which it leads us to expect. [But i]n this sense, a solipsist might say that physics is true; for although he would suppose that the sun and the moon, for instance, are merely certain series of perceptions of his own, yet these perceptions could be foreseen by assuming the generally received laws of astronomy. . . . A man who, without being a solipsist, believes that whatever is real is mental, need have no difficulty in declaring that physics is 'true' in the above sense, and may even go further and allow the truth of physics in a much wider sense. This wider sense, which I regard as the more important is as follows: Given physics as a deductive system, derived from certain hypotheses as to undefined terms, do there exist particulars or logical structures composed of particulars, which satisfy these hypotheses? If the answer is in the affirmative, then physics is completely 'true.'

The view that we know as "Russell's structuralism" and the considered view of *AoM* both assume that physics is capable of being "completely true" and not just capable of anticipating our perceptions. In Russell's considered view, the claim that physics is "completely true" requires the logical construction out of percepts and other events of the primitive entities of physics. But these constructions rest on assumptions that run counter to his "structuralism," since they assume that some of the relations which order percepts *also* order events which are not percepts.

In opposition to Russell, Ramsey maintains that the narrower sense of "truth" suffices for our understanding of the truth of physics: Physics is true just in case it enables us to anticipate what we perceive to occur. But such anticipations depend only on a *partial* interpretation of physics, and this is precisely what is afforded by Campbell's dictionary. The claim that this sense of "truth" is sufficient is what is "in exact contradiction to Russell" because it contradicts his demand for an interpretation in terms of logical constructions capable of securing the "complete" truth of physics.

A point that Ramsey made fully explicit only in "Theories" is that we are justified in rejecting logical constructions because an argument from physics to an observation sentence—to a truth of the primary system—is recoverable from a representation of physics in the form of (1) *provided* we include in *F both* the theoretical claims of physics *and* a Campbellian dictionary relating theoretical and observational vocabulary. Once we recognize this as the condition of adequacy which an explication of the notion of *truth* appropriate to physics requires, it remains only to show that its satisfaction has no need for the complex system of constructions of Russell's notion of interpretation. And, of course, the Ramsey sentence of a partially interpreted theory *does* satisfy this condition of adequacy.

This interpretation of the fragment is difficult to resist, if only because it renders Ramsey's proposal so entirely in keeping with the nature of his other modifications of Russellian doctrines. It is especially reminiscent of the most celebrated among them, namely, his simplification of the type hierarchy of the first edition of *Principia*. That modification does not eliminate the need for logical constructions, but it greatly reduces their number while also simplifying their definition. It, too, rests on the supposition that a weaker condition of adequacy suffices, one according to which the success of a theory of classes is judged by its ability to recover certain known truths of mathematics, while resolving only what Ramsey distinguished as the mathematical paradoxes to which the logical notion of class gives rise.

The historically surprising upshot of our discussion is that the idea of representing physical theories in the schematic form of a Ramsey sentence emerged from Ramsey's critical reaction to the program of logical analysis that Russell set forth in *AoM*. It is of course also true that Ramsey's reflections could not have taken the path they did without the notion of a partial interpretation, which emerges in this discussion of Russell and which appears to have been suggested to him by Campbell's notion of a dictionary.

Ramsey's justification of the adequacy of his positive view depends on the thesis that the truth of a theory consists in its successful anticipation of truths formulated in the primary system. This explicit equation of truth with predictive success relative to the primary system is entirely consistent with his later formulation of the task of "Theories" (Ramsey 1929, p. 212): "Let us try to describe a theory simply as a language for discussing the facts the theory is said to explain." The only alternative conception of truth appropriate to theories that the fragment considers is Russell's. It is, however, instructive that Ramsey's rejection of Russell's conception of "complete truth" targets only *AoM*'s particular implementation of it in terms of logical constructions. Nowhere does the fragment critically address Russell's general idea that there is more to the truth of a theory than its successful anticipation of what we will observe. Nor does Ramsey address whether the thesis that truth is more than predictive success can be adequately formulated within the framework of his proposed reconstruction: Working under the assumption that he had hit upon the right notion of truth for physics, he likely saw no need to do so.

Stathis Psillos (2006, pp. 82–85) has proposed a very different account of Ramsey's discussion of Russell's schematic representation of theories, one according to which Ramsey anticipated Lewis's recognition of the importance of natural or real properties and relations: Psillos argues that Russell's schematic representation of physics "looks very much like a Ramsey sentence. But unlike Russell, Ramsey did *not* adopt a structuralist view of the content of theories." According to Psillos, Ramsey, unlike Russell, understood the range of the variables bound by the new existential quantifiers to be restricted to *real*—or, as Psillos also says in order to stress the connection with Lewis, *natural*—properties and relations: "Ramsey takes theories to imply the existence of definite (or real) relations and properties. Hence, it's no longer trivial (in the sense explained above) that if the theory is empirically adequate, it is true. His Ramsey sentences can be seen as saying that there are *real properties and relations* such that. . . ." (Psillos 2006, p. 84).

Psillos's interpretation of Ramsey hinges on this point. But Ramsey's fragment is entirely explicit about the basis for his rejection of the adequacy of (1)—as he is about his own alternative proposal in terms of partial interpretation—and nowhere in his discussion of theories does Ramsey even mention a view to the effect that the range of the variables should be restricted to real or natural properties and relations. To address this apparent difficulty for his interpretation, Psillos turns to Ramsey's paper "Universals" (1925) where Ramsey explicitly distinguishes between those propositional functions ϕx in which ϕ occurs as a "name" and those in which it occurs as an "incomplete symbol." Ramsey argues that in the former case, ϕ can "stand by itself" as the name for an "object," while in the latter, it can be seen to disappear on analysis. Psillos sees in Ramsey's distinction between those properties which have names and those which are necessarily represented by incomplete symbols an anticipation of the distinction between natural and nonnatural properties. But as we will see, Ramsey's distinction, in "Universals" between two kinds of property does not support this interpretation. In "Universals" it is important for Ramsey to show that in atomic or elementary propositions we have only "combinations of objects" which are indifferent to the distinction between particulars and universals. Some properties (Ramsey writes "functions") *must* be expressed by propositional functions and cannot be named, while others need not be expressed by propositional functions. Ramsey argues that the distinction between these two cases is ignored by mathematical logic, because it treats all properties as merely a means to classes and so represents them all by propositional functions; in so doing, it makes the distinction between universals and particulars appear more compelling than it is, and as is well known, it is this distinction which "Universals" is concerned to challenge.

To illustrate his distinction between these two kinds of property, Ramsey considers the complex property, bearing R to a or S to b, and argues that it cannot be expressed by a "simple symbol" or "name," such as ϕ, but demands a variable; thus $\phi x = xRa$ or xSb, which, as Ramsey says, "explains not what is meant by ϕ itself but that followed by any symbol x it is short for xRa or xSb" (1925, p. 130). But not all properties must be expressed by propositional functions, and those that need not be so expressed can serve as the constituents of atomic propositions. In advancing this claim, Ramsey is not raising an objection to propositional functions like xRa or xSb because they are merely extensional or because they fail to pick out the neutral objectual constituents of elementary propositions. Nor is there anything evidently "unnatural" about such a property: It might be the property of being causally connectible to a by a massive particle, or to b by a massless particle—a perfectly reasonable property of events in the context of Minkowski space-time. Ramsey's point is that by focusing on such functions for our conception of a property, we are distracted from seeing that the distinction between particular and universal that is fostered by the notion of a propositional function may not hold at the level of elementary propositions. As such, his isolation of properties that can be named—and that therefore do not require a propositional function for their expression—is at best a contribution to our understanding of the logical form of elementary propositions, one with no obvious bearing on a property like that of bearing R to a or S to b along the dimension of naturalness or reality.

To recapitulate, Ramsey's departure from Russell on the notion of interpretation appropriate to physics consists of four principal differences:

(1) Ramsey, unlike Russell, holds that it is only necessary to show that physics is true in the narrower of the two senses Russell distinguished, namely, the sense that corresponds to the successful anticipation of our perceptions (empirical adequacy). And by contrast with Russell, Ramsey says very little about the domain over which theoretical claims are interpreted, except to exclude interpretations which would render the theory "tautological."

(2) For Ramsey, the philosophical interest of a reconstruction of physics consists *wholly* in its ability to represent *all* of our reasoning with the theory to statements in the primary system. As we have since come to learn, the Ramsey sentence has a distinguished position among reconstructions which dispense with theoretical vocabulary. Unlike a Craig transcription of a theory, the Ramsey sentence does not just have the same observational consequences as the theory: It is an immediate consequence of its logical form that although the Ramsey sentence is strictly weaker than the theory, it is satisfiable if and only if the theory is satisfiable. And since the Ramsey sentence of a theory preserves the interconnections among primary parameters which depend on their association with the theory's secondary parameters, it is capable of representing our reasoning with the theory to a degree that is unmatched by a Craig transcription.

(3) Ramsey departs from Russell in not holding that the domain of an interpretation of a physical theory must consist of logical constructions out of percepts and other events which "have spatiotemporal continuity with percepts." Russell had argued that unless its theories can be given such an interpretation, the empirical character of physics is compromised. Ramsey dispenses with Russellian constructions in favor of a partial interpretation, which he argues suffices to explicate the empirical character of physics. This is perhaps his most fundamental departure from Russell. For Ramsey, theories are abstract Hilbertian schemes which acquire their empirical character through their association with our perceptions by the inclusion of a dictionary. But the notion of a dictionary is precisely what is missing from Russell's schematic representation of physics. This is a major break, not just with Russell, but also with the whole logicist framework of Russell and Whitehead.

(4) Ironically, when understood in accordance with his considered view, it is Russell who is not a structuralist. Ramsey, believing that he had simplified the task of interpretation and relieved it of the cumbersome logicist assumptions which so deeply informed Russell's approach, fully assimilated the structuralism exhibited by the subsequent development of the partial interpretation reconstruction into his own account of theories. The difficulties associated with that reconstruction are obscured in Ramsey's presentation only because his view is developed within a framework that equates the truth of physics with its empirical adequacy.

1.6 Constructive Empiricism and Partial Interpretation

Van Fraassen's constructive empiricism is explicitly formulated in a "semantic framework," according to which theories are conceptualized not as linguistic objects, as the logical empiricists had supposed, but as classes of models. And by contrast with antirealist reconstructions like Ramsey's, constructive empiricism explicitly endorses the distinction between a theory's truth and its empirical adequacy. Indeed, this distinction is essential to one of the main arguments that has been marshaled in support of the view over and against scientific realist alternatives to it. But despite its many noteworthy differences from logical empiricism, constructive empiricism is subject to an objection that bears comparison with the difficulties which confront the partial interpretation reconstruction of theoretical knowledge.[27]

In his original formulation, van Fraassen based constructive empiricism on a central tenet and a definition that has its source in his development of the semantic conception of theories. The central tenet is that it is always more rational to accept a theory which postulates unobservable entities as *empirically adequate* than it is to believe it to be *true,* and the definition is that to *accept a theory as empirically adequate* is to hold that the observable phenomena can be "fitted into" a model belonging to the class of models that comprises the theory. In his most recent work, van Fraassen has emphasized the importance of a theory's "empirical grounding" rather than its empirical adequacy. I will discuss this notion of empirical grounding in some detail in a later section.[28] But for the present, my focus is empirical adequacy and constructive empiricism's conception of how empirically adequate and true theories differ.

Van Fraassen replaces the rough formulation of the explication of empirical adequacy just cited with a more precise formulation: A theory is *empirically adequate* if all the phenomena involving observable entities—"the appearances"—are isomorphic to the "empirical substructures" of one of its models, where the empirical substructures are those parts of the models of the theory that have been set aside "as candidates for the . . . representation of observable phenomena." As van Fraassen (1980, p. 64) expresses it:

> To present a theory is to specify a family of structures, its *models;* and secondly, to specify certain parts of those models (the empirical substructures) as candidates for the direct representation of observable phenomena. The structures which can be described in experimental and measurement reports we can call *appearances;* the theory is empirically adequate if it has some model such that all appearances are isomorphic to empirical substructures of that model.

Three features of this explication are especially worth noting. The first is the emphasis on structures rather than languages, on empirical sub*structures* rather than observational sub*languages*. The second is that empirical substructures are characterized by the observability of what their constituents represent; in this respect, van Fraassen's approach retains a fundamental similarity to that of the logical empiricists by addressing issues of epistemic access from an empiricist perspective that stresses the epistemic priority of observation. The third is the characterization of empirical adequacy by the isomorphism of empirical substructures with appearances—the structures consisting of observable things and events—rather than by the truth of a distinguished set of sentences. This last feature explains why the explication is presented as an improvement on an earlier characterization, according to which "a theory is empirically adequate exactly if what it says about the observable things and events in this world is true" (van Fraassen, 1980, p. 12). The earlier account fails to reflect the fact "that the explication of [empirical adequacy] . . . is intimately bound up with our conception of the structure of a scientific theory" (ibid.), which, as van Fraassen explains, consists in a theory's structure as a class of models rather than as a "syntactic" or linguistic object. The proposed explication remedies this failing by basing the definition of empirical adequacy on structures and the relations among them.

The isomorphism which empirical adequacy requires is one which holds between the relations of a model of the theory and relations among entities which—just because they are observable—are unproblematically accessible to us. The emphasis on empirical substructures, and the characterization of empirical adequacy by isomorphism, represent departures from the logical empiricists' account of theories in terms of theoretical and observational *vocabularies* and the several *languages* these vocabularies generate. These departures are essential to van Fraassen's program of supplanting the language-dependent conception of theories, which he takes to have been the major obstacle to a successful characterization of empirical adequacy in the logical empiricists' account of theories.

It is therefore worth noting that if we set to one side the issues separating their different explications of empirical adequacy, van Fraassen's notion of acceptance, in the sense of the acceptance of a theory as empirically adequate, is basically the same as Carnap's notion of acceptance that we quoted earlier (in note 9 of the Introduction):

> For an observer X to "accept" the postulates of T, means here not simply to take T as an uninterpreted calculus, but to use T together with specified rules of correspondence C for guiding his expectations by deriving predictions about future observable events from observed events with the help of T and C. (Carnap 1956a, p. 45)

And indeed, it will emerge from our discussion that the differences between constructive and logical empiricism are much less fundamental than proponents of constructive empiricism have supposed.

The explication of empirical adequacy receives considerable attention in *The Scientific Image*—and rightly so, given the difficulties that had been urged against earlier empiricist characterizations of the notion in terms of the observation vocabulary of a theory and the sentences that are generated from it.[29] But although *The Scientific Image* has a great deal to say about the proper explication of empirical adequacy, it does not contain a comparably careful explication of the *truth* of what a theory which goes beyond the appearances "says about unobservable things and events"; nor is there even a statement of the conditions such an explication should satisfy. It is only clear that the explication of truth for theories which go beyond the appearances should extend the one given for empirical adequacy to cover both what a theory tells us about appearances and what it tells us about unobservable entities. A proposal that satisfies this desideratum without appealing to any notions other than those employed in the explication of empirical adequacy—except that of things and events which transcend observation—is to say that an empirically adequate theory involving unobservable entities is true if a model of the theory which witnesses its empirical adequacy can be extended to one such that *all* phenomena—both observable and unobservable phenomena—are isomorphic to that model.[30] This explication is not only appropriately semantic in its representation of theories as classes of models and its characterization of their truth in terms of isomorphism, it also preserves a feature of constructive empiricism which it shares with the partial interpretation reconstructions of the logical empiricists: In the case of appearances, isomorphism is ostensibly "anchored" by the fact that appearances are structures which are generated by relations which hold between entities that are epistemically accessible to us by observation. By contrast with our access to appearances, *The Scientific Image* leaves our epistemic access to unobservable entities and the relations among them just as indefinite as we found it to be on the partial interpretation account of theoretical knowledge. But as we will now show, this omission undermines constructive empiricism's use of the *central tenet* to establish its superiority to scientific realism.

To understand why an account of our access to theoretical domains is needed—even on a view which, like constructive empiricism, seeks to undermine our confidence in what we believe we know of the structures they constitute—recall that constructive empiricism is presented as an alternative to scientific realism that is free of reductionist commitments. Constructive empiricism does not deny that theories include representatives of unobservables among the elements of their domains; it is merely cautiously agnostic regarding theoretical claims about unobservable phenomena. It concedes "that physical theories do indeed describe much more than what is observable, but [holds that] what matters is empirical adequacy, and not the truth or falsity of how they go beyond observable phenomena."[31] Now by the *central tenet,* constructive empiricism is alleged to be preferable to any of its scientific realist alternatives, because it is the more rational position: It accepts some theories as empirically adequate, but it refrains from the stronger and unwarranted supposition that they are true. But on what would seem to be its own understanding of the truth of empirically adequate theories which go beyond the appearances, it can be shown that there is no substantive difference between believing a theory true and accepting it as merely empirically adequate. This is contrary to the claim that when measured against scientific realism, constructive empiricism is the more rational position to adopt toward theories which go beyond appearances.

To establish this observation, it suffices to show that any theory which is empirically adequate according to constructive empiricism must contain among the models that comprise it one that isomorphically represents both the observable, and the unobservable, phenomena. But we know from our discussion of the model-theoretic argument—a discussion which effectively places partial interpretation reconstructions in the context of the semantic view of theories—that to show this, it suffices that the domain of one of the theory's empirically adequate and partially abstract models is in one-one correspondence with the domain of observable and unobservable entities. Using the construction deployed in the model-theoretic argument, an empirically adequate theory can then be transformed into a true one, since the construction ensures that if there is a model belonging to the theory that contains an empirical substructure isomorphic to the appearances, the theory contains a model that isomorphically represents the phenomena involving *both* observable and unobservable entities. But this suffices to show that, according to constructive empiricism, the theory is not merely empirically adequate, but true. This undermines the use of the tenet that it is always more rational to accept a theory as empirically adequate than to believe it true, because it shows that for constructive empiricism, the difference between an empirically adequate theory and a true one—and thus the basis for favoring constructive empiricism over scientific realism—rests entirely on an assumption about cardinality. Since it is obvious that the rationality of this choice should not hinge on a question of cardinality, it follows that the combination, constructive empiricism plus van Fraassen's development of the semantic view of theories, is an inherently unstable position.

Van Fraassen has subsequently acknowledged that the use of isomorphism, which characterized his earlier formulations, was uncritical:

> [I]n *The Scientific Image* constructive empiricism was presented in the framework of the semantic view of theories. . . . See for instance Chapter 3 section 9, p. 64 where I define empirical adequacy using unquestioningly the idea that concrete observable entities (the appearances or

phenomena) can be isomorphic to abstract ones (substructures of models). (Van Fraassen 2008, pp. 385–386)

It is clear from this passage that, according to van Fraassen's diagnosis, the problem with his earlier formulation concerned the explication of *empirical adequacy* and the fact that isomorphism was improperly applied to *concrete* observable entities and *abstract* structures. But this misses the key point: *The Scientific Image* fails to explicitly provide an explication of what it means for an empirically adequate theory that goes beyond the appearances to be *true,* and the natural extension of its explication of empirical adequacy to address this omission rests on a vacuous use of isomorphism. This use is itself a consequence of constructive empiricism's failure to address our epistemic access to unobservable entities. Constructive empiricism's understanding of the truth of theories in terms of the isomorphism of structures is therefore just as problematic as the partial interpretation account of theoretical claims. Just as there is nothing in the partial interpretation account to exclude the use of a construction according to which the truth of theoretical claims reduces to their satisfiability in an arbitrary extension of a model of the true observation sentences, so also, there is nothing in constructive empiricism's understanding of the truth of an empirically adequate theory to exclude the use of this same construction to extend a model which witnesses the theory's empirical adequacy to one which witnesses its truth. On both views, the truth of an empirically adequate theory involving unobservable entities—however empirical adequacy is explicated—is insufficiently distinguished from its satisfiability. This consequence is not the result of a mere oversight of either approach. Both views are invested in programs that understate the basis for our confidence in the representation of reality to which we are guided by physical theories. But once our epistemic access to domains that transcend observation is clarified, neither the skepticism about unobservable phenomena that is characteristic of constructive empiricism, nor the deflationist view of the controversy between realism and instrumentalism that has been so pervasive among advocates of partial interpretation, is defensible.

The inadequacy of both partial interpretation and constructive empiricist reconstructions of theoretical knowledge derives from their common failure to articulate a plausible account of our epistemic access to theoretical domains. In fact, by contrast with the logical empiricist tradition it seeks to replace, constructive empiricism shows little appreciation of the need to even address this issue. Since its inception, constructive empiricism has been focused on showing how, when formulated within the semantic conception of theories, an empiricist framework can resolve the difficulties which beset earlier explications of empirical adequacy.[32] But the issues which need to be addressed are more substantial than this and concern the scope of the empiricist framework which constructive and logical empiricism both share. That framework requires a fundamental reevaluation; and when measured against the difficulties to which it is subject, the questions, "Are theories linguistic objects or families of structures?" and "Is empirical adequacy a property of observation sentences or of empirical substructures?" pale by comparison.

It is interesting to note as a final corollary to the foregoing discussion that the possibility of transforming an empirically adequate theory into a true one poses a difficulty for van Fraassen's agnosticism regarding unobservable entities that it does not pose for Carnap's metaphysical neutrality regarding theoretical claims. This is because it is incoherent to be agnostic about a theory's truth while accepting it as empirically adequate if an empirically adequate theory can always be transformed into a true one. Rightly or wrongly—which is not presently at issue—Carnap is *neutral* between accepting a theory as empirically adequate and believing it to be true; he can therefore embrace an argument that calls into question the distinction between empirically adequate and true theories, since the failure of this distinction actually supports his claim that the question "Realism or instrumentalism?" is one of practical decision, rather than one whose answer involves theoretical or factual knowledge.[33] Hence, when taken in conjunction with his metaphysical neutrality, Carnap's reconstructions—whether his partial interpretation, or Ramsey-sentence reconstructions. or the account based on Hilbert's epsilon operator—are all capable of sustaining a challenge to the distinction between truth and empirical adequacy in a way that the signature agnosticism of van Fraassen's constructive empiricism is not.[34]

The burden of this section has been to show that there are two questions about the nature of theoretical knowledge that have not been adequately addressed by any of the accounts we have considered: (1) What is the methodology by which physics gains epistemic access to theoretical domains? (2) How are the claims of physics about such domains distinguished from the claims of pure mathematics? The central idea behind the proposal I will develop in Chapter 2 is that the theoretical claims of physics are distinguished by the fact that they purport to be about properties and relations—more generally, parameters—that are empirically well founded by robust theory-mediated measurements. The well foundedness of its parameters makes the theoretical domains of physics epistemically accessible; it accounts for the specificity of the theoretical claims of physics, and it distinguishes its claims from those of pure mathematics. But the methodology of robust theory-mediated measurements involves considerations that have been largely missed by the accounts of empirical adequacy and predictive success that have dominated the partial interpretation and closely related traditions. How the application of theory-mediated measurements to domains which transcend observation developed, and how it bears on our questions, is the subject to which we now turn.

2 Molecular Reality

2.1 The Molecular Hypothesis

The molecular hypothesis is the existential claim of the molecular-kinetic theory of heat that gases and liquids are made up of large numbers of extremely small constituent particles which are in relative motion and capable of acting on one another, with the nature of these "molecular constituents" left unspecified and open to further study.[1] This hypothesis forms the constructive core of the molecular-kinetic theory and is the basis for its reconstruction of familiar thermodynamic parameters. The principles of classical mechanics, together with two key statistical assumptions (the equipartition of energy and the Maxwell-Boltzmann distribution law for velocities), form the theory's principle-theoretic components.[2] The hypothesis is deliberately formulated very generally and with few specific commitments. It is not even necessarily committed to the character of the dynamical interactions among the molecular constituents of gases and liquids; these may involve elastic collisions—as was supposed in the earliest formulations of the kinetic theory and as is suggested by the fact that the Brownian motion never stops—but the hypothesis is as flex-

ible on this point as it is on many questions of molecular structure, including even the shapes of molecules.

To the extent that a justification of the molecular hypothesis
has been supposed to be bound up with the fortunes of the
molecular-kinetic theory, the soundness of the hypothesis has
been thought to be seriously compromised by the problems confronting that theory.[3] But this is a mistake. The molecular hypothesis is detachable as the theory's constructive-theoretic component and is susceptible to a justification that is capable of surviving
difficulties—even insurmountable difficulties—that attach to the
principles of that theory. Indeed, the general correctness of the
molecular-kinetic theory is not even essential to the arguments
which sought to "derive" the molecular hypothesis from the phenomenon of Brownian motion.[4] Both Einstein and Perrin—the
two principal proponents of the molecular hypothesis on whom
we will focus—were careful to argue that the considerations that
establish molecular reality do not depend on special hypotheses
about the nature of molecules which, given the then-current state
of knowledge, were often arbitrary.[5] Nor do they require the un-
restricted validity of the principles of the kinetic theory: The idea
that the molecular hypothesis might be true was correctly regarded as compatible with granting only restricted validity to
the kinetic theory's principle-theoretic components, all of which
required extensive revision with the discovery of what ultimately
proved to be quantum-mechanical phenomena.

2.2   Molecular Reality and Brownian Motion

Einstein's 1905 paper "On the Movement of Small Particles Suspended in a Stationary Liquid Demanded by the Molecular-Kinetic
Theory of Heat" contains just a passing reference to Brownian

motion. Einstein remarks only that "[i]t is possible that the movements to be discussed here are identical with the so-called 'Brownian molecular motion'; however, the information available to me regarding the latter is so lacking in precision, that I can form no judgment in the matter" (Einstein 1905, p. 1). It is clear from his presentation of his results that Einstein saw as the main novel contribution of his paper not the proof of the existence of molecules—which he appears even at this early date to have taken for granted—but the description of the motion of particles which, though small, are several orders of magnitude larger than molecules. The 1905 paper dealt with the translational motion of such particles, and another, published a year later, extended the account to their rotational motion (Einstein 1906). That paper explicitly addressed Brownian motion, as its title ("On the Theory of Brownian Movement") makes clear. Einstein emphasized that should his account prove correct, it would establish a striking difference between the molecular-kinetic theory of heat and classical thermodynamics, since it would demonstrate that the former theory allows for violations of the second law of thermodynamics, even by particles of visible size.[6] Einstein also observed that the law he derived for the description of the motion of these small but visible particles could be used to estimate the values of many molecular quantities, among them Avogadro's number and the molecular diameter.[7]

The explanation afforded by the development of the molecular hypothesis of what we now know as Brownian motion goes well beyond the qualitative idea that the pollen grains first observed by the Scottish botanist Robert Brown are bombarded on all sides by the molecules of the stationary liquid in which they are suspended. It does so by exploiting the consequences of the notion that the Brownian particles and the molecules have the same mean kinetic energy. However, it was far from obvious when Einstein wrote that an explanation of Brownian motion based on the ideas of the molecular-kinetic theory could succeed. The objections were both conceptual and empirical. On the conceptual side, the cytologist Karl von Nägeli argued that the effect of the molecular bombardment on a pollen grain would be to leave the grain stationary and quivering about a fixed position, and could not therefore be invoked to account for its characteristic random motion.[8] And Felix Exner's attempt to experimentally determine the *speeds* of Brownian particles produced values that were too low to be remotely compatible with the assumption that the average kinetic energy of Brownian particles is the same as the theoretically required value of the average kinetic energy of the molecules of the liquid in which they are suspended.

Einstein's theoretical analysis demonstrated that the motion of Brownian particles is not only compatible with the molecular-kinetic theory but can be fully described within its theoretical framework. His theory's focus on the *mean square displacement* of a Brownian particle rather than its speed was especially prescient since, unlike the particle's speed, this quantity can be empirically determined; and the law which Einstein derived for it, according to which the displacement increases in proportion to the time elapsed after the pattern of a random walk, can be shown to be in accordance with experimental results.[9]

In the introduction to his exposition of the experimental basis for Einstein's theory,[10] Perrin explained why the speed of a Brownian particle was the wrong parameter for Exner to have attempted to measure, and he showed how underestimating the extraordinary complexity of such a particle's actual trajectory— it approximates a continuous but nowhere differentiable curve— undermined Exner's experimental results. Perrin not only confirmed Einstein's theoretical analysis of Brownian motion; he also succeeded in setting a course that would eventually lead to the demonstration of the molecular hypothesis, and he did so with a comprehensiveness and decisiveness that had eluded earlier work. What was it about Perrin's contribution that warrants its being accorded the status of a turning point in the assessment of the molecular hypothesis?

There is a subtle difference between questions of methodology when they are raised in the context of an empirical investigation like Perrin's and when they are raised in connection with mathematics. In the nineteenth century the central philosophical question posed by mathematical knowledge was whether the mathematical reasoning on which it is based relies on some notion of "Kantian intuition." By the twentieth century this question had been transformed into one about the *finitary* character of mathematical reasoning, as emphasized particularly by Hilbert. Without in any way disparaging the conceptual interest of the study of novel proof techniques in mathematics, the question of the finitary character of mathematical proof can be addressed, at least initially, without engaging the mathematical innovations that characterize historically important proofs. Not only does the question of the finitary character of mathematical proof have an independent philosophical and methodological interest, but as the history of the subject demonstrates, it yielded to a purely logical analysis, culminating in Gödel's theorem on the completeness of first-order logic. The further analysis of the concept of finitism focused on limitations of formal systems, and this led to many deep and surprising results—beginning with Gödel's theorem on the incompleteness of first-order arithmetic.* It is, how-

---

*Editor's note: The notion of "completeness" has different senses in the two cases. Gödel's initial theorem says that all first-order logical truths can be demonstrated in a single recursively enumerable and thus finitary axiomatization, although contingent sentences are not logical truths and thus are indemonstrable in such an axiomatization.

ever, far from obvious that a similarly general strategy can illuminate the methodological issues that confront us in the case of empirical theories.

The reconstructive framework of the partial interpretation view of theories which appeal to unobservables is an attempt to extend the logical analysis that proved so useful in the case of mathematical proof by dividing the vocabulary of a theory into its observational and theoretical components and proposing that the evidentiary basis of such an empirical theory is expressible in its observation language. Within this framework, the natural condition of adequacy for an empirical theory is the derivation of all and only the true statements that are expressible in the observation language of its partial-interpretation reconstruction. We saw earlier that this approach gives rise to certain difficulties in connection with its account of the epistemic status of theoretical claims and the existence claims associated with them. The suggestion that a different approach is needed is borne out by the considerations which led Perrin and his contemporaries, including the notable skeptics among them, to become convinced of the real possibility that a demonstration of the truth of the molecular hypothesis—insofar as one can speak of demonstrations in empirical science—was within their grasp. The considerations Perrin put forward to show that the molecular hypothesis conforms to experience clearly had a justificatory force that earlier attempts at establishing molecular reality lacked. Even if we retain some reservations about the decisiveness that Perrin's reasoning was accorded, it seems essential to an understanding of

---

Gödel's incompleteness theorem concerns axiomatizations of first-order arithmetic, and the negative result is that not all arithmetical truths have finitary demonstrations in a single such axiomatization (provided that the axiomatization is consistent): There are thus always indemonstrable arithmetical truths.

the methodology by which we assess theories like the molecular-kinetic theory—and especially the existential hypotheses they contain—that we should understand what it was about his reasoning which accounted for its favorable reception.

Assuming that Perrin and his contemporaries were consistent in the application of the methodological standards to which they subscribed, we can immediately exclude an account of Perrin's success that is based on the hypothetico-deductive method or the method of inference to the best explanation. The problem with these accounts is that they fail to distinguish Perrin's reasoning from the metaphysical atomists of the seventeenth and eighteenth centuries who preceded him and whose defense of their existence claims regarding atoms was widely regarded as fundamentally inadequate. In this connection, editions of Wilhelm Ostwald's textbook on physical chemistry prior to its fourth edition and his 1907 *Monist* article are often cited. But the source of dissatisfaction with earlier forms of atomism is already implicit in an observation of Newton's: "For if the possibility of hypotheses is to be the test of the truth and reality of things, I see not how certainty can be obtained in any science."[11] Here, Newton's phrase, "the possibility of hypotheses," should be understood as an allusion to hypothetico-deductive reasoning—so common among the corpuscularean philosophers—and the fact that such reasoning is capable of capturing only the consistency of a hypothesis with experiment.

Perrin brought to the debate about theories of atomic reality—and indeed to the debate about all theories which purport to describe a reality that is hidden from observation—a higher standard of what is required by their conformity to experience than is captured by either of these accounts. How is this higher standard reflected in his argument for molecular reality?

Perrin's remarks on the concordance of various methods for determining the value of Avogadro's constant and on the vertical concentration of granules in a stationary fluid have attracted a great deal of attention in the recent philosophical literature. Two passages in particular have been frequently cited in support of a probabilistic understanding of Perrin's intended justification of the molecular hypothesis. It has also been argued that a probabilistic reconstruction of these passages explains why, whatever his intentions, Perrin's contribution to securing the molecular hypothesis *should* be seen as carrying the conviction it does. Since these passages are also central to my proposal for understanding the nature of Perrin's defense of molecular reality, it is necessary that we should consider them in some detail. The passages clearly show Perrin making an informal appeal to probability, but, as I hope to show, they do not support a probabilistic understanding of his argument for molecular reality, and they are incapable of identifying what it is that justifiably made his contribution the turning point in the assessment of the molecular hypothesis that it was.

The first of these passages concerns the concordance of various determinations of the value of Avogadro's constant:

> Our wonder is aroused at the very remarkable agreement found between values derived from the consideration of such widely different phenomena. Seeing that not only is the same number obtained by each method when the conditions under which it is applied are varied as much as possible, but that the numbers thus established also agree among themselves, without discrepancy, for all the methods employed, the real existence of the molecule is given a probability bordering on certainty.[12]

To spell this out just a bit more, Perrin's point is not merely that closely agreeing determinations of Avogadro's constant N are recoverable from a variety of empirically determinable parameters which are functionally related to it. The functional relations these different determinations are based on have their source in an extensive variety of different branches of physics; and their agreement extends to the determination of N by experimental methods and functional relationships that are independent of one another and are independent in particular of those determinations of the constant based on the kinetic theory.[13] This last point is one to which we will return when we come to Thomson's discoveries. Perrin's suggestion is that what ties all these different determinations together—what explains their concordance—is the central idea of the molecular hypothesis, namely its interpretation of Avogadro's constant as a measure of the *number* of molecules in a specified volume. Whether or not the extensive agreement in the measured values of N is *by itself* sufficient to constitute an argument for molecular reality, or to explain the force that argument commands,[14] it certainly carries prima facie conviction—a conclusion Perrin expresses with the assertion that it confers on the molecular hypothesis "a probability bordering on certainty." But since the quoted passage occurs near the very end of a monograph devoted to expounding all the extant contributions of Perrin and other investigators to the vindication of the molecular hypothesis, it is not unreasonable to suppose that the full argument for molecular reality is more intricate than this summary statement suggests. Until we have explored this possibility, it would be premature to conclude that Perrin's argument is adequately represented by a probabilistic reconstruction.

There is a second passage that has attracted the attention of probabilistic reconstructions; it occurs in the context of Perrin's discussion of his vertical concentration experiments. After describing the results of his measurements of granules showing that the "granular energy" is independent of the masses of the granules, the differences in the density of the individual granules relative to the density of the medium, and independent as well of the rapidity with which the concentration of the granules increases with a fall in height, Perrin remarks:

> I do not think this agreement [with the hypothesis that the mean kinetic energy of the granules is identical with the theoretically predicted value of the mean kinetic energy of the molecules of the liquid in which they are suspended] can leave any doubt as to the origin of the Brownian movement. To understand how striking this result is, it is necessary to reflect that, before the experiment, no one would have dared to assert that the fall of concentration would not be negligible in the minute height of some microns, or that, on the contrary, no one would have dared to assert that all the granules would not finally arrive at the immediate vicinity of the bottom of the vessel. The first hypothesis would lead to a value of *zero* for N′, while the second would lead to the value *infinity*. That, in the immense interval which *a priori* seems possible for N′, the number should fall precisely on a value so near to the value predicted, certainly cannot be considered as the result of chance.[15]

It is important to observe that in this passage, N′ is the quantity 3RT / 2W where T is the absolute temperature of the liquid, R is the gas constant, and W is the mean kinetic energy of the granules. N′ will indeed turn out to be equal to N—to Avogadro's

constant. But at this stage of the argument, where Perrin is re-
porting the results of his vertical concentration experiments, N'
is merely a quantity that is functionally related to the mean
kinetic energy of the granules, without necessarily being the measure of a *number* of anything, still less, a number of molecules.

There is a particular type of probabilistic reconstruction of this
and other of Perrin's remarks that I wish to focus on: namely,
those reconstructions which take Perrin's argument for molecular
reality to be based on a comparison of the molecular hypothesis
with its negation. Psillos has extensively developed a reconstruction along these lines. Commenting on the second of the two passages, he writes:

> Perrin became immediately convinced that "this agree-
> ment can leave no doubt as to the origin of Brownian
> movement. . . . [A]t the same time it becomes very difficult
> to deny the objective reality of molecules." What con-
> vinced him, he . . . says, was that on any other hypothesis
> (better, on the negation of the atomic hypothesis), the ex-
> pected value of N' from the study of the movement of
> granules suspended in a liquid would be either infinite or
> zero—it would be infinite if all granules actually fell to the
> bottom of the vessel, and zero if the fall of the granules
> was negligible. Hence, on the hypothesis that matter is
> continuous, the probability that the predicted value of N'
> would be the specific one observed would be zero; on the
> contrary, this probability is high given the atomic hypoth-
> esis. This, Perrin noted, "cannot be considered as the re-
> sult of chance."[16]

Psillos evidently sees Perrin's argument as showing that his experiments rule out the negation of the molecular hypothesis—

which is here equated with the hypothesis that matter is continuous—by showing it to have a probability near zero. Psillos does
not say that *Perrin* explicitly claimed to rule out the negation of
the molecular hypothesis, but he does see his remarks as "almost"
claiming it.[17]

It is easy to see why it would be desirable to be able to show
that Perrin pursued the strategy Psillos attributes to him, since it
would certainly differentiate Perrin's contribution from less compelling justifications of molecular reality based on hypothetico-
deductive reasoning. There are however serious difficulties with
a probabilistic reconstruction that targets the negation of the molecular hypothesis. There is the interpretational difficulty that
Perrin does not mention the negation of the molecular hypothesis but focuses on his analysis of the distribution of the concentrations of granules in a liquid. In the course of his exposition,
Perrin certainly appeals to our naïve expectations: No one would
be prepared to deny that the differences will be found to be very
slight in the concentrations of granules belonging to thin vertically distributed layers that are only a few microns apart; and similarly no one would be prepared to deny that all the granules will
eventually settle near the bottom of the vessel. But the point of
calling attention to these appeals to naïve expectations is to highlight a feature of the motion of the *granules* that is evident only
when it is subjected to a close experimental analysis. It is not
merely that the Brownian particles appear to gross observation
to be in perpetual motion: Their concentration exhibits a *fine
structure* which, over the course of a few hundred microns, displays a rarefaction profile which matches the one that the atmo-
sphere exhibits only over six kilometers.

Each of the several different types of granule Perrin studied
varies in concentration according to a geometric progression.
Most importantly, while the concentration of the granules in a

thin layer becomes increasingly dense the closer the layer is to the bottom of the vessel, the concentration remains stable at every level. Without the guidance afforded by his experimental investigations, only the extreme values for N′ suggest themselves. Nearly identical concentrations in closely situated layers would yield a value of 0 for N′, and should all the granules eventually settle near the bottom of the vessel, this would yield a value of infinity for N′. So far as the evaluation of probabilistic reconstructions is concerned, Perrin's allusion to our expectations about the concentration of granules shares an important point of similarity with his remarks on concordance.

Although he is not, in either of these cases, addressing how we should bet on the assumption that the molecular hypothesis is false, he is remarking on how limited our expectations are in its absence: Without the molecular hypothesis—or as Perrin says, on the basis of what "*a priori* seems possible"—it is remarkable that in the interval between zero and infinity, the molecular-kinetic theory should predict the observed value of N′. But Perrin is also calling our attention to the surprising nature of what his experimental analysis of Brownian motion reveals about the rarefaction profile of the Brownian particles. This is a phenomenon that is striking *whatever* its correct explanation.

The assertion that the molecular hypothesis is the only *known* hypothesis that leads to correct expectations about Brownian motion is of course much weaker than the assertion that it is the *only* hypothesis that enjoys this property. Evidently nothing we have said is directed against the soundness or the persuasiveness of the methodological practice of eliminating known alternatives. On the contrary, the considerations we have raised support it, since an argument that addresses our expectations in the *absence* of the molecular hypothesis rests on nothing more than our recognition that *no other hypothesis known to us* provides the right answers to these questions. Unlike a reconstruction which traces the persuasiveness of Perrin's reasoning to its successful elimination of the negation of the molecular hypothesis, the exclusion of known alternatives is not committed to what might be demanded by hypotheses regarding which there is no possibility of our forming a definite judgment. For this reason, any such reconstruction obscures rather than illuminates Perrin's reasoning while raising seemingly intractable problems of its own.[18]

There may be an interesting probabilistic justification of the molecular hypothesis. But if it is made to rest on our ruling out the negation of the molecular hypothesis, it is not Perrin's justification. Moreover, Perrin was right not to have pursued such a strategy, since the nature of our expectations in the absence of the hypothesis is clear in a way that our expectations on the basis of its negation are not. Involving as it does a quantification over all possible theories, to argue on the basis of the negation of the molecular hypothesis can hardly be assimilated to reasoning with a surveyable collection of alternatives of the sort that are familiar in the case of the event spaces of probability theory. Not only is such reasoning very different from reasoning which appeals only to explicit contrary hypotheses, we have no idea how to expand the negation of the molecular hypothesis into an exhaustive disjunction of comparably explicit alternative contrary hypotheses. This is why a probabilistic reconstruction of Perrin's strategy along these lines fails to be convincing as an account of why his justification was rightly accorded the importance that it was. In short, if the method of hypothesis is based on a conception of the evidential support for the molecular hypotheses that is too weak, this reconstruction of Perrin's method of argument promises a justification that, by any measure that can reasonably be applied to the justification of an empirical hypothesis, is too strong.

2.3   The Nature and Status of Perrin's "Connecting Link"

There is a tendency to take Perrin too literally and to approach his overview of his experiments with the expectation of finding a direct argument for the molecular hypothesis. This tendency is encouraged by Perrin's own exposition, which intersperses the description of his findings with conclusions about molecular reality and its role in the explanation of Brownian motion. But reading Perrin in this way conceals rather than clarifies the structure of his overall argument. That argument is indeed an argument for molecular reality.[19] But it has a subtlety that is easily missed.

In order to appreciate Perrin's contribution it is necessary to recognize that, as striking as Brownian motion is to gross observation, its fine-structure is more striking still, and even its precise *description* was a considerable theoretical and experimental achievement. Perrin's argument for molecular reality is a development of the following observation: *modulo the fact that the granules do not influence one another, even when they approach each other as closely as one granular diameter, their motion constitutes* a visible model *of the molecular-kinetic theory.* Perrin's experiments demonstrate this in precise detail, more specifically, they show:

> (G) *The mean kinetic energy of the granules is the same at any given temperature regardless of the character of the granules or the character of the liquid in which they are suspended.*

This is a conclusion drawn at the level of the various kinds and sizes of visible *granules* involved in his experiments.

But Perrin's main goal is to secure a premise which affords what he calls a "connecting link" between the mean energies of the granules and the molecules of the emulsion in which they are suspended:

(I) *The mean kinetic energy of the* molecules, *whose inter-action with the granules—whether by contact or by some more complicated force—is held to be responsible for their motion, is the same as the mean kinetic energy of the* granules.

There are essentially three considerations which Perrin advances on behalf of (I). The first is a plausibility argument derived from van 't Hoff's investigations of osmotic and gaseous pressure, and the extension to liquids (dilute solutions) of the molecular-kinetic theory of gases to which they led. Van 't Hoff's theoretical investigations show:

(M) *At the same temperature all the molecules of all fluids have the same mean kinetic energy, which is proportional to the absolute temperature.*

In particular, this holds as we consider molecules of *larger and larger size,* and it suggests the possibility that it might be generalized further still:

Let us now consider a particle a little larger still, itself formed of several molecules, in a word a *dust.* Will it proceed to react towards the impact of the molecules encompassing it according to a new law? Will it not comport itself simply as a very large molecule, in the sense that its mean energy has still the same value as that of an isolated molecule? This cannot be averred without hesitation, but the hypothesis at least is sufficiently plausible to make it worth while to discuss its consequences. (Perrin 1910, p. 20)

Perrin's formulation is somewhat misleading. When he writes that "the hypothesis is sufficiently plausible to make it worth while to discuss its consequences," he does *not* mean that the only basis for (I) is the plausibility argument which is indicated by van 't Hoff's theory. The theoretical path to (I) is complemented by his experimental investigations in two ways. First, while the theoretical argument converges on (I) by the consideration of larger and larger molecules, Perrin's experimental results converge on it by the investigation of *smaller and smaller granules*. But second, (I) receives further support from the fact that theory-mediated measurements of molecular parameters that are based on considerations altogether different from Perrin's—because they do not appeal to his connecting link involving the mean energy of the molecules—yield essentially the same results for the values of these parameters as his own.

If one focuses only on the theoretical discussion which motivates (I), one will miss the full force of Perrin's argument for molecular reality because the basis for this key premise will have been dramatically understated. This, it seems to me, is precisely where van Fraassen goes wrong in his account of Perrin.[20] In a discussion of the paragraph of Perrin's we have been considering, van Fraassen writes:

[W]e should note that the theoretical derivations which Perrin assumes are largely dependent also on assumptions added to the kinetic theory, in the construction of specific models. . . . The addition Perrin made to this already

almost century-old story follows the same pattern. As
Achinstein emphasizes, Perrin also introduces an addition
to the theory, a "crucial assumption, viz. that visible particles comprising a dilute emulsion will behave like molecules in a gas with respect to their vertical distribution"
(Achinstein 2001, p. 246). Note that this is a blithe addition: Perrin argues for its plausibility, but in terms that
clearly appreciate the postulational status of this step in
his reasoning. . . . On [its] basis, the results of measurements made on collections of particles in Brownian motion give direct information about the molecular motions
in the fluid, always of course within the kinetic theory
model of this situation. (van Fraassen 2009, p. 20)

Van Fraassen (apparently following Achinstein) appears to be
suggesting that Perrin's connecting link is a crucial but *mere* assumption—"a blithe addition" to the molecular-kinetic theory—
which has the same status as the simplifying assumptions about
the shapes of molecules that enabled earlier theoretical calculations of molecular parameters. But the plausibility argument
which is indicated by van 't Hoff's theory does not exhaust Perrin's justification for his connecting link; and one of the things,
as we will see, which separates Perrin's determination of molecular parameters from many of his predecessors is the care with
which he *avoids* the use of the simplifying assumptions which
are characteristic of the theoretical calculation of molecular-
parameter values. Although it is not a point that Perrin stresses,
if we put to one side the correct *explanation* of Brownian motion,
it is possible to arrive at Perrin's conclusions about its character
without at any stage invoking the molecular hypothesis or the
molecular-kinetic theory.

To recapitulate, for Perrin it is the combination of two sets of considerations that supplies the initial justification for the premise that the mean kinetic energy of the molecules is the same as the mean kinetic energy of the granules. First, there are those considerations that derive from the extension of the molecular-kinetic theory to liquids; and second, there are those that derive from his own work with Brownian motion. The connecting link that they support is of course central to Perrin's whole discussion, since it is on its basis that he is able to infer of the molecules such characteristics as their number and their masses, estimates of their sizes and, in general, the values of many parameters that had been without empirical determination. As Perrin put the point in *Atoms:*

> If the agitation of the molecules is really the cause of the Brownian movement, and if that phenomenon constitutes an *accessible connecting link* between our dimensions and those of the molecules, we might expect to find therein some means for getting at these latter dimensions. This is indeed the case, and we have moreover a choice of methods we may employ. I shall discuss first the one that seems to me the most illuminating.[21]

But Perrin's justification for his connecting link is not exhausted by what is theoretically mandated regarding molecules of increasing size and his extrapolation from what he discovered to be true of granules of different sizes. The link receives further support from the fact that theory-mediated measurements of molecular parameters that are based on altogether different considerations yield essentially the same molecular-parameter values.

2.4   Perrin's Argument for Molecular Reality

The application of the molecular hypothesis to the explanation of Brownian motion begins with the recognition that, given our understanding of the dynamical behavior of continuous media, it is not possible to locate the source of the granules' motion so long as we adhere to the assumption that the liquid in which they are suspended conforms to our conception of a dynamical system with the structure of a continuous fluid. This observation is followed by the demonstration that the experiments which reveal the character of Brownian motion can be used to exploit the phenomenon as a "means for getting at molecular dimensions" since they show it to provide a "connecting link" which allows us to empirically determine molecular parameters when we extrapolate from the mean kinetic energy of the granules to the mean kinetic energy of the molecules of the liquid.

The granular concentration experiments and the experiments involving the translational and rotational displacement of the granules employ different methods and experimental techniques as well as different functional relationships among the relevant parameters. Despite the variety of assumptions on which the experiments are based, they agree in the values they disclose for the various *granular* parameters, and they are mutually supporting of one another and of the conclusions to which they lead about *molecular* parameters. Moreover, the findings based on the extrapolation to the mean kinetic energy of the molecules are a source of information about molecular reality and its empirical determination that are susceptible to increasing refinement in a way that theoretical calculations of molecular parameters based on the classical kinetic theory of gases are not. Those calculations rest on simplifying assumptions, which, like the perfect spherical shape of molecules, are at best a convenience. As Perrin remarks in his Nobel Lecture (1926, p. 10)

> The accuracy of [my] determinations [of molecular parameters], so far of several hundredths, can certainly be improved: the same does not apply to values obtained from the kinetic theory of gases, because here perfecting the measurements would not diminish the uncertainties inherent in the simplifying assumptions which were introduced to facilitate the calculations.

This point is completely missed by van Fraassen when he writes:

> To be realistic we should note that the theoretical derivations which Perrin assumes are largely dependent also on assumptions added to the kinetic theory, in the construction of specific models. Most of the work proceeds with models in which the molecules are perfect spheres, for example, though Perrin (1910, p. 14) notes that other hypotheses are needed in other contexts. As long as the simple models work, to allow a transition from the empirically obtained results to values for the theoretical parameters, and as long as these values obtained in a number of different ways agree with each other and with what is theoretically allowed—to within appropriate margins of error—this counts as success. (van Fraassen 2009, p. 20)

But in the passage from *Brownian Movement and Molecular Reality* to which van Fraassen refers, Perrin is discussing the assumptions that go into the theoretical calculations, which, in his

Nobel Lecture, he is concerned to *contrast* with his own theory-mediated determinations of molecular parameters.

With this perspective in mind, let us turn to a somewhat fuller reconstruction of Perrin's argument for molecular reality. Writing in 1888 Louis-Georges Gouy successfully excluded virtually all the extant alternative explanations of Brownian motion that sought to locate its cause in a source external to the fluid in which the particles are suspended.[22] (It was clear that the explanation of Brownian motion cannot reside in the particles themselves, since they are observed not to influence one another even when they pass as close as one diameter.) After recounting the achievements of Gouy and his predecessors, Perrin (1910) remarks in his section 5 on how microscopic observation of a container into which a "continuous fluid" has been poured, and some very small colored particles ("colored powders") have been placed, shows that the "coordinated" (i.e., nearly equal and parallel) motions of the colored particles gradually gives way to "decoordinated" or randomly distributed motions of the particles. After a short time, in order to find a region in which the motions of the particles are coordinated, it is necessary to consider smaller and smaller portions of the fluid. But when the motion of the fluid appears to have ceased, and we imagine it to be in a state of static equilibrium, a microscopic examination reveals the incessant motion of the colored particles and shows the equilibrium we thought we had observed to have been illusory.

Closer examination of the motion of the colored particles shows it to exhibit a remarkable property: Over time, if in a small region the motion of some of the particles stops, it is immediately compensated by motion in another small region where it is taken up by the particles of that region. Perrin calls this phenomenon "recoordination" and argues that it shows the dissemination of motion from one region to another to proceed not "vertically" from a small region to its smaller and smaller nested subregions, but "horizontally" from a small region of the fluid to another comparably small region which is disjoint from it. If the fluid were indeed continuous, very general constraints that guide our conception of the motion of a continuous medium would require that the dissemination should be wholly vertical. The fact that the motion is not disseminated vertically but horizontally strongly suggests, again on the basis of very general considerations—in this case, principles governing elastic or nearly elastic collisions among particles—that the fluid has a particulate structure, with the particles of the medium imposing a bound on the vertical dissemination of the motion.

Perrin's discussion of the source of the phenomenon of Brownian motion illustrates the utility of the distinction between "principles" governing dynamical processes and "constructive" hypotheses about the constitution of the fluid by dialectically confronting characteristics of the principle-theoretic components of diverse theories in order to arrive at a decision about a constructive hypothesis concerning the fluid's constitution. Reflection on the phenomenon of Brownian motion and those general dynamical principles appropriate to particle motion and the motion of continuous media then leads us away from one theory of the composition of the fluid and toward another.[23]

Perrin's argument from Brownian motion to molecular reality continues with the observation—made precise by the analysis of Einstein and others[24]—that according to the kinetic theory small but visible particles should exhibit the features that the theory attributes to molecular motion. Perrin's observations of granules in Brownian motion dramatically confirmed this prediction. But Perrin also showed how, on the assumption that the motion of the granules is the effect of their interaction with the molecules of the liquid in which they are suspended—and, in particular, on the assumption that the mean kinetic energy of the molecules is the same as the mean kinetic energy of the granules—it is possible to devise empirical determinations of parameters that had previously resisted such determination.

As noted in Section 2.3, Perrin's justification for the identification of the mean kinetic energy of the molecules with that of the granules is based on his experimental demonstration that the mean kinetic energy of the granules is independent of their size and the earlier theoretical investigations of van 't Hoff extending the kinetic theory of gases to liquids. As Perrin demonstrated, the identification has a special importance since it carries with it epistemic access to molecular reality. But unlike the role of a correspondence rule, which associates a "theoretical" parameter with an "observable" one, Perrin's connecting link has nothing to do with the explanation of our understanding of "mean kinetic energy" when this term is applied to unobservable entities. It is a precondition of his analysis that the term is one we understand, and it is taken for granted that our understanding of it does not change merely because it is applied to particles too small to be visible. What is new is Perrin's use of the mean kinetic energy and the connecting link in which it occurs to address the empirical determination of parameters that qualify molecules. From this perspective, the central novelty of Perrin's contribution to our assessment of the molecular hypothesis was his demonstration that Brownian motion could be exploited to serve this purpose.

After Perrin, the molecular hypothesis stood in marked contrast to both the metaphysical atomists and their inability to secure anything comparable in connection with the properties of atoms, and the ether theorists and their failure to empirically determine such theoretically mandated properties as velocity relative to the ether. The importance of the fact that molecular parameters are empirically well founded by theory-mediated measurements does not require that the estimated values to which such measurements lead are the "true" values of the parameters; what is important is that the methodology by which such parameter values are obtained should be susceptible of systematic refinement by a well understood methodological practice.

The methodological significance of Perrin's achievement was quickly recognized by Poincaré who remarked that in light of Perrin's work there is no denying that we "see" atoms.[25] This remark should be understood in terms of an analogy of Perrin's: Just as the infrared is an extension of visible light, so also the empirical determinability of molecular quantities shows how a combination of experimental and theoretical reasoning can be regarded as an extension of our faculty of vision, an extension in the sense that it exhibits the same possibilities of greater resolution and refinement.[26] As Perrin (1926) noted in his Nobel lecture quoted earlier in this chapter, his methods allow for the possibility of systematically refining the measurement of molecular parameters in a way that theoretical calculations based on idealizing assumptions do not. It is this feature of Perrin's contribution to our epistemic access to molecular reality that supports a comparison with perception and distinguishes it from the hypothetical reasoning of the early atomists. And it is this departure from the earlier atomistic tradition that attracted Poincaré's attention and elicited his remark.

It is common to suppose that the philosophical question the molecular hypothesis raises for realism is how decisively Perrin's investigations of Brownian motion establish the truth of the molecular-kinetic *theory*. But there are at least two difficulties

with this way of framing the issue of realism and Perrin's possible contribution to our understanding of it. First, Perrin's experiments could not possibly establish the molecular-kinetic
theory simply because the difficulties with the theory were too
great. At most, the Einstein-Perrin analysis of Brownian motion
can be regarded as having shown the limited validity of some of
its basic principles, such as the equipartition principle. Second,
it can reasonably be argued that the defense of an existence claim
like the molecular hypothesis does not proceed by establishing
the truth of the theories in whose explanations the hypothesis
forms an integral and indispensable part. The considerations on
which the molecular hypothesis depends are rather: (i) the difficulty of finding the *source* of Brownian motion if the fluid in
which the granules are suspended is assumed to have a continuous structure, (ii) the strength of Perrin's connecting link trans-
ferring the mean kinetic energy of the granules to that of objects
of smaller dimensions, and (iii) the justification of the empirical
well-foundedness of the parameters which characterize molecules.
Even though the molecular-kinetic *theory* was highly problematic, the developments occasioned by the analysis of Brownian
motion are more than sufficient to have set a course which culminated in the justification of the molecular *hypothesis*.

The empirical determinability of any number of parameters of
the molecular-kinetic theory cannot of course show the theory
to be true or even empirically adequate. But it can be decisive for
the justification of the existential claim on which the theory rests.
Although this is a significantly weaker claim than the conclusion
that the molecular-kinetic theory is not merely empirically adequate but true, it is clearly favorable to the realist conviction that
science is capable of capturing salient truths about reality, even a
reality that transcends our experience. For it tells us that the

empirical determinability of molecular parameters is capable of establishing molecular reality even if it fails to justify with comparable conviction theories about that reality. In this respect our use of empirical well-foundedness goes further than the view—recently advanced in van Fraassen (2009)—that something akin to what we have described as theory-mediated measurements serve only to show that the molecular-kinetic theory is "empirically grounded." I say "akin to theory-mediated measurements," because van Fraassen is concerned with a special case of the bearing of the measurement of certain parameters on *theories*—not with the general phenomenon of theory-mediated measurement. The question van Fraassen is addressing is: What is the nature of the support that a variety of concordant measurements of different parameters can bestow on a *theory* when it is the *theory itself* that guides the implementation of the measurement of these parameters? The conclusion for which he argues is that the support that a theory receives in such a case is different from the support that attaches to it on the basis of its predictive success.

Provided the parameters and the theory meet certain conditions,[27] such measurements are capable of showing the theory which informs them to be empirically grounded. This is *less* than empirical adequacy because it allows for the possibility that there are appearances which the theory does not save. But it is also *more* than empirical adequacy, since a predictively successful theory might well fail to be empirically grounded. And most importantly for van Fraassen, empirical groundedness, like empirical adequacy, falls far short of truth.

As we noted earlier, Perrin observed that his results are susceptible to increasing refinement—unlike purely theoretical calculations of parameter values that are based on idealizations which cannot be relaxed without undermining the calculation. Recall that the theory-mediated measurements carried out by Perrin determine *granular* parameter values. They proceed within the framework of the molecular kinetic theory in the sense that they exploit functional relationships between parameters belonging to this framework. None of these measurement results depend on the molecular hypothesis. The transition to *molecular* parameters is based on Perrin's extrapolation to the identity of the mean molecular kinetic energy of the granules with that of the molecules comprising the fluid in which they are suspended. The justification for this extrapolation is based on Perrin's demonstration of the constancy of the mean molecular kinetic energies of granules of increasingly smaller size, and the concordance of the determination of molecular parameter values employing this assumption with their determination by methods based on considerations that are independent of the framework of the molecular-kinetic theory.

The conclusion Perrin's measurements support is not merely the well-groundedness (in the sense of van Fraassen) of the molecular-kinetic *theory*, but the *truth* of the molecular *hypothesis*. This point appears not to have been assimilated by either the realist or antirealist side of the debate in the recent literature on Perrin and Brownian motion. It is certainly true that the considerations advanced by Perrin do not contribute to our confidence in the truth of the molecular-kinetic theory; but this was never their point. Perrin showed how, by detaching the question of molecular reality from the truth of this theory, it is possible to undermine skeptical views of molecular reality that appeal to the fact that the theories that assume it are sometimes highly problematic. Given the formulation of a multiplicity of clear and independent criteria for empirically determining the properties of molecules—with all the criteria concordant, and the empirical determination of the relevant parameters therefore robust—what more could one reasonably require of a demonstration of their existence? Only that comparable success in the exploration of the epistemic accessibility of the submolecular and subatomic levels should also be forthcoming—something that was dramatically borne out by the subsequent development of the subject.[28]

By way of summary, Perrin's argument for molecular reality has five stages. The first stage infers the particulate nature of a fluid from the phenomenon of Brownian motion and the principles that are characteristic of the motion of particles and the motion of continuous media. This is not yet an inference to an *explanation* of Brownian motion, let alone an inference to its *best* explanation. At this stage, the aim is, as Perrin puts it, to show how the phenomenon of Brownian motion might be said to "logically suggest" the particulate nature of the fluid. The second stage of Perrin's argument is directed at securing the basis for extrapolating from our experience with granules of various sizes to the value of the mean kinetic energy of the much smaller particles which, on the molecular hypothesis, constitute the fluid. The third stage consists in the exploration of the consequences of the connecting link, argued for at an earlier stage, for the empirical determination of a host of molecular parameters. The fourth stage consists in recounting the support that the connecting link receives from the remarkable uniformity and concordance of the determination of parameter values to which it leads with various other determinations of these parameter values. The final stage infers from what the earlier stages have revealed the explanation of Brownian motion in terms of the molecular hypothesis.

As Perrin points out, it would be a mistake to view his argument for molecular reality as the simple exploration of the empirical success of the predictions to which we might be led if we were to begin with the molecular hypothesis.[29] In light of our account it is clear why we should reject such a reconstruction of Perrin's argument. First, it leaves out the critical role of the confrontation of the phenomenon of Brownian motion with our understanding of the motion of continuous media. Second, proceeding "hypothetically" is compatible with the use of theoretical parameters for which we lack any principled means for empirically determining their values. Third, reasoning "hypothetically" makes no provision for the critical importance of the separate requirement—dramatically shown to be satisfied by the methods Perrin employed—that the various determinations of molecular parameters should be concordant. Fourth, hypothetical reasoning is not *essentially* coupled with empirical methods that carry with them principles by which they can be indefinitely corrected and refined. All of these aspects of Perrin's reasoning are overlooked by hypothetico-deductive reconstructions of it. And, at the very least, the second and fourth are ancillary to an account that represents Perrin's methodology as an instance of inference to the best explanation. The molecular hypothesis may well be the best explanation of Brownian motion, but its true epistemological support is far more nuanced than such a characterization suggests.

2.5   Thomson and the Constitution of Cathode Rays

There are many investigations that complement Perrin's conclusions about molecular reality by bringing to bear evidence that is wholly unrelated to Brownian motion. An important and representative one among them occurred in connection with J. J. Thomson's discovery of the corpuscular nature of cathode rays, representing what we would today describe as the discovery of the electron, the first of the subatomic particles of modern physics.[30] Thomson's discoveries were contemporaneous with Perrin's and are a necessary extension of the molecular hypothesis to the atomic and subatomic levels. In addition to extending the molecular hypothesis into the region of the subatomic, Thomson's investigations led to a determination of N based on Townsend's discovery that in the process of electrolysis the charge per ion is the same as the charge per hydrogen molecule. This determination of N does not depend on knowledge of the mean kinetic energy of the molecules under study but derives it from knowledge of the charge on the electron. It vindicates the robustness of Perrin's results by yielding a determination of N that is entirely independent of Perrin's but is in sufficiently close agreement for the two determinations to count as concordant.[31]

Thomson's (1906) Nobel Lecture, "Carriers of Negative Electricity," sets out with remarkable clarity the way in which an intricate combination of experimental and theoretical considerations led to the discovery of the corpuscular nature of cathode rays and the determination of the value of e.[32] His argument for the particulate nature of electricity shares the general theory-mediated-measurement structure of Perrin's for molecular reality.

The initial consideration that Thomson advances in favor of the corpuscular nature of cathode rays is a species of argument by *analogical extension* from our familiarity with the behavior of charged particles under the influence of electric and magnetic forces. The novelty of Thomson's contribution to this analogical argument consists in his elicitation of the requisite behavior from cathode rays under the influence of an electric force by placing them in an environment (vacuum tubes—Thomson's "highly exhausted tubes") in which the presence of gases is negligible so that the rays are not insulated from the electric field. Given our gross experience with the motion of charged objects such as pith balls, the behavior of cathode rays when subjected to magnetic and (under appropriate conditions) electric forces is suggestive of the idea that the rays are composed of corpuscles. But Thomson did more than elaborate an argument by analogy. He went on to develop a series of arguments, closely similar in their methodological significance to those employed by Perrin in his empirical determination of parameters relevant to the molecular hypothesis; with this contribution, Thomson, like Perrin, established much more than the mere plausibility of an existential hypothesis.

Having determined the velocity of cathode rays, Thomson turned his attention to the determination of e/m, the charge-to-mass ratio of the particles that compose them. As Thomson explains, this ratio is central to his argument since its proper determination is one of the key components of his response to Hertz's discovery that cathode rays are capable of penetrating gold leaf, a discovery that presented a serious difficulty for the view that cathode rays are composed of corpuscles. The difficulty arises from the fact that since hydrogen atoms, the smallest corpuscles then known, are not capable of penetrating gold leaf, electrons would have to be several orders of magnitude smaller than even the hydrogen atom.

The claim that electrical particles are capable of passing through "even the densest matter" can already be found in the writings of Benjamin Franklin. Thomson's achievement was to establish their capacity to pass through gold leaf by an independent assessment of their size, rather than by invoking an assumption about the electron mass that would enable the corpuscular hypothesis to meet the difficulty raised by Hertz's discovery. The procedure Thomson used for determining that electrons are vastly smaller than hydrogen atoms depends on first learning the value of e/m, and then inferring, on the basis of knowledge of e, the mass of the electron.

Thomson describes a method for determining e/m that exploits the behavior of cathode rays under the influence of an electric field. A brief theoretical analysis shows the determination of e/m to depend on an easily measured displacement of a patch of phosphorescence where the cathode rays strike the glass tube. It is a remarkable fact, which Thomson emphasizes, that, using this method, one finds the same value for e/m however the rays are produced, and whatever the velocity of the particles that compose them—provided only that this velocity does not come close to that of light. The value of e/m is found to be independent as well of the nature of the electrodes and of the gas environment of the tubes containing the rays. Thomson (1906, pp. 149–150) also remarks on the great variety of other conditions—conditions not involving cathode rays—under which electrons can be obtained: by heating in the case of some substances, and by cold the case of others; by alkali metals when these are exposed to light; and by radioactive substances, in which they are given out in large quantities. In all of these cases e/m has the same constant value.

But although the soundness of Thomson's argument for the value of the electron mass and much else depends fundamentally on the correctness of his determination of e/m, Thomson nowhere supposes that the constancy of e/m is the basis for inferring the particulate nature of cathode rays.[33] Thomson's determination that the charge-to-mass ratio is approximately 1,700 times the charge-to-mass ratio for the hydrogen atom raises the question of whether its value is a consequence of the smallness of the mass of the electron compared with that of the hydrogen atom or of the fact that the charge on the electron is much greater than that on the hydrogen atom.[34] Thomson seeks to establish the first alternative, since a vastly smaller m would address Hertz's discovery.

The reasoning on the basis of which Thomson infers a very small value for m depends on a number of subarguments devoted, ultimately, to the determination of e. Thomson's subarguments proceed from a variety of experimentally established premises, with the choice of premise guided by its ability to be combined with a theoretically established connection between measured quantities and the quantities whose values are being sought. The subarguments are mutually supporting: Some are invoked to relax a restriction in the scope of one or another earlier argument, while others establish the same conclusion in a different experimental setting, thereby lending greater generality to the argument. The proof that the large value of e/m is attributable to the small value of the electron mass is developed using C. T. R. Wilson's discovery that a charged particle forms a nucleus around which water vapor condenses and forms drops. Appealing to Stokes's Law of Fall for the determination of the average size of the water droplets enclosing electrons, Thomson arrived at an estimate of the value of the charge on each carrier of negative electricity. And as Thomson notes, a simpler method, deployed by H. A. Wilson, yields the same estimate.

In his pursuit of a more precise determination of e, Millikan later noted many places where Thomson's assumptions could be questioned.[35] The points on which Millikan laid particular stress were Thomson's reliance on Stokes's Law of Fall and the difficulty one encounters when repeating Thomson's experiments of making it "even approximately true that each droplet contain[s] only a single unit of charge" (Millikan 1917, p. 50). Millikan's determination of the absolute value of e on the basis of his experiments with oil drops significantly improved on Thomson's results. But important as Millikan's investigations were in showing that the apparent unitary value of e was genuine and not merely a statistical phenomenon, and that charges increase or decrease by unitary steps, its revelation of the need for significant improvements and corrections to Stokes's Law and to Thomson's experimental procedures and results did not undermine Thomson's demonstration of the corpuscular nature of cathode rays. Rather, it extended and corrected the approximative reasoning which guided Thomson's determination of the electron's fundamental properties. In doing so, Millikan confirmed the incisiveness of Thomson's general methodological conception.

The contribution of Thomson is distinguished by its isolation and adept experimental investigation of the phenomena which form the basis of the analogical extrapolation to the particulate nature of cathode rays. This hypothesis is made compelling by the way in which an initially plausible analogical argument is elaborated by considerations that are drawn from a variety of independent sources and then used to refine our determinations of the values of a number of fundamental parameters with what can justifiably be judged as increasing accuracy and precision. Most importantly, Thomson, like Perrin, showed how otherwise seemingly inaccessible quantities are empirically determinable by a variety of independent methods, all of which accord with one another. These arguments are not demonstrative and are incapable of precluding the possibility of skepticism about the reality of the unobservable entities under study.

However the credibility of a skeptical assessment of Thomson's existential hypothesis demands that we focus on the analogical component of the argument to the exclusion of its technical elaboration by theory-mediated measurements. Such a focus misses what is distinctive about both Thomson's and Perrin's contributions, since it is precisely their articulation of methods for the robust empirical determination of the parameters that qualify the entities they were investigating that separates their work from the just-so stories of an earlier atomist tradition and removes any legitimate doubt about the correctness of their existence claims. It was precisely his appreciation of this feature of modern atomism that informed Poincaré's mature view of the status of the molecular hypothesis.

3 Poincaré on the Theories of Modern Physics

3.1 Poincaré on "True Relations"

Poincaré's discussion of what he refers to as "true relations"—relations in the form of functional relations connecting various physical parameters—occurs in the chapter of *Science and Hypothesis* entitled "The Theories of Modern Physics." The discussion is guided by two observations: (1) physical theories are often *indifferent* to the nature of the constitution of the things whose behavior they seek to describe; (2) physics thrives on the discovery of relations which yield various different but concordant determinations of physical parameters; these are the true relations Poincaré wishes to highlight. We will see how these two observations inform Poincaré's remarks on modern physical theories, and in particular, how the emphasis he eventually came to place on true relations led him to change his view of the hypotheses about which physics should—and should not—be indifferent. In addition to being supported by the relevant texts, the account of Poincaré I will propose allows us to understand very clearly why he found Perrin's research on Brownian motion, together with certain other discoveries which accompanied it, so decisive for the demonstration of molecular reality.

Those of Poincaré's remarks about the primacy of relations that derive from (1) may have been partly suggested by his familiarity with Newton's discovery that for many astronomical calculations it suffices to consider just the centers of mass of the bodies whose motions we seek to describe. This is not a philosophical thesis but a mathematical theorem that depends on demonstrable properties of the relevant forces, and it evidently allows us to set entirely to one side the complexities which questions of the constitution of bodies raise. In this case, our ability to ignore the "nature" of the objects under study does not rest on a general view about their opacity to further study. But although the example of Newton may have been importantly suggestive, it is clear that it does not exhaust the considerations which led to Poincaré's emphasis on the *indifference* of physical theory to the constitution of the things whose dynamical behavior it seeks to describe.

Poincaré cites three examples to illustrate the thesis that the abiding import of physical theories consists in their ability to direct our attention to true relations. The first example concerns the isolation and characterization of periodic phenomena. We are in possession of a characterization of periodic systems that is general and abstract in the sense that it is independent of the mechanical or electrical nature of the periodic process to which it can be applied. Its discovery was facilitated by theoretical principles of energy conservation and least action, and the characterization captures both the periodicity of the motion of a pendulum and the periodicity of an electric current. Poincaré cites this example to stress that the recognition of the periodic character of an electric oscillation is an important advance whatever our view of the source of such oscillations. The example illustrates the fact that theoretical advances often abstract overspecific commitments to the nature and constitution of the physical systems and the origin of the phenomena to which they can be successfully applied. Of the three examples we will be considering, this one comes closest to supporting a "structuralist" interpretation of him: periodicity is a property that is susceptible to a purely abstract mathematical characterization. But there is a large gap between the perception that physics benefited from the trend toward abstraction that characterized mathematics toward the end of the nineteenth century and the thesis that our knowledge of the material world is restricted to its mathematical or structural properties.

Poincaré next cites an example which illustrates the possibility that diverse and even mutually incompatible theoretical starting points can agree in the relationship they find between two physical parameters, namely, the relation between absorption and anomalous dispersion.[1] Of this example, Poincaré writes:

> Numerous theories of dispersion have been proposed. . . . But, what is remarkable, is that all the scientists who came after Helmholtz reached the same equations, starting from points of departure in appearance very widely separated. I will venture to say that these theories are all true at the same time, not only because they make us foresee the same phenomena, but because they put in evidence a true relation, that of absorption and anomalous dispersion. What is true in the premises of these theories is what is common to all the authors; this is the affirmation of this or that relation between certain things which some call by one name, others by another. (Poincaré 1902, authorized Halsted translation, p. 141)

Poincaré's stress on the agreement in equations is not an abstract metaphysical claim about the importance of "pure form" or structure. As Poincaré notes on the immediately preceding page, the point of the equations in which he is interested is to express relations; and in the present case, the equations on which everyone since Helmholtz converged express the relation between the known quantities of absorption and anomalous dispersion.

The sense of this passage is seriously distorted in the unauthorized translation, reprinted in the 1952 Dover edition, where, on p. 162, we read:

> But the remarkable thing is, that all the scientists who followed Helmholtz obtain the same equations, although their starting-points were to all appearances widely separated. *I venture to say that these theories are all simultaneously true; not merely because they express a true relation—that between absorption and abnormal dispersion.* In the premises of these theories the part that is true is the part common to all: it is the affirmation of this or that relation between certain things, which some call by one name and some by another.

The sentence I have italicized omits altogether Poincaré's contrast between foreseeing phenomena and isolating true relations, and it fails to make clear that what Poincaré is concerned to isolate as the common part of these different approaches is the fact that the equations they share express the same relation between absorption and anomalous dispersion. Without this qualification it is easy to misinterpret the sentence with which the passage concludes, and to suppose that the parameters whose relation the equations express might be represented purely abstractly by variables of the appropriate type. Poincaré's point is not this, but is rather the physically interesting observation that despite the fact that the nature of light and the refractive medium were differently conceived in different theories, all of them nonetheless converged on the relation between absorption and anomalous dispersion.

But my primary interest is Poincaré's last example, since it is most closely related to our study of the shift in his evaluation of the molecular hypothesis. It involves the case of a problematic theory, the classical kinetic theory of gases, leading to the nineteenth-century discovery of the correct relation between osmotic and gaseous pressure. Poincaré's point is that the relation to which the example draws our attention retains its interest and importance whatever the status of the molecular model that suggested it:

> The kinetic theory of gases has given rise to many objections, which we could hardly answer if we pretended to see in it the absolute truth. But all these objections will not preclude its having been useful, and particularly so in revealing to us a relation true and but for it profoundly hidden, that of the gaseous pressure and the osmotic pressure. In this sense, then, it may be said to be true. (Poincaré 1902, authorized Halsted translation, p. 141)

The background to Poincaré's point may be explained as follows.

When a solution of sugar in water is separated from a pure solvent—such as water—by a membrane that allows water but not sugar to pass, then water forces its way through the membrane and into the solution. This process results in greater pressure on the solution side of the membrane; this pressure is osmotic pressure. Once it was known how to measure osmotic pressure, there arose the question of how to determine its relation to the concentration and temperature of the solution. This was a nontrivial problem which led eventually to extending the kinetic theory of gases to include liquids, a development that figured prominently in Perrin's argument for his connecting link (as we saw in Section 2.3). The key to its resolution was van' t Hoff's observation that with sufficiently dilute solutions the osmotic pressure is the same as the pressure which the dissolved substance would exert as a gas.

Although the identification of gaseous and osmotic pressure is readily suggested by transposing the model of gas pressure (as the impact of gas molecules on the sides of a container) to the collisions of the molecules of the solution with a semipermeable membrane, its justification did not require this hypothetical picture, but was made compelling by focusing on the behavior of the relevant parameters without appealing to their atomistic interpretation. Thus it was first discovered that the relation between osmotic pressure and the volume of the dissolved substance—sugar in our example—at a fixed temperature satisfies Boyle's equation, $pV =$ constant. Then it was discovered that the constant in the relation between pressure and concentration varies with temperature in accordance with Gay-Lussac's law, $pV/T =$ constant, and that this constant is independent of the nature of the solvent and the dissolved substance. Finally it was recognized that this constant can be so represented that it has a value very close to that of the universal gas constant. On the basis of these considerations it was concluded that *"the osmotic pressure is exactly the same as the gas pressure which would be observed if the solvent were removed, and the dissolved substance were left filling the same space in the gaseous state at the same temperature."*[2] It is also possible to express the general significance of the identification of osmotic and gaseous pressure as a discovery that stands whatever the fortunes of the hypothetical model: Every mole of any nonelectrolytic dissolved substance in a dilute solution has the same characteristic energy regardless of the nonelectrolytic substance or the nature of the solvent.

There are three lessons that can be drawn from the example of osmotic and gaseous pressure. First, if we focus on the argument for their identification that is independent of the molecular model of gases and liquids, it is clear that its appeal to the fact that osmotic pressure, like gaseous pressure, satisfies the equations of Boyle and Gay-Lussac, is not in aid of the idea that our knowledge of these parameters, insofar as it rests on their inclusion in certain equations, is "wholly structural." The value of these equations does not consist in their expression of mere "mathematical forms" which are shared by gases and liquids, but in what they express about the relations between the known physical parameters which enter into them. Second, reservations about the identification of osmotic and gaseous pressure had nothing to do with the nature of our knowledge of pressure but were concentrated entirely on the molecular model of gases and liquids. The reasons for such reservations were many, but certainly an important one among them was the fact that a commitment to this model is not needed to motivate the identification. Third, the principal significance of the discovery of the identity of gaseous and osmotic pressure is that it enables the empirical determination of the relation of osmotic pressure to many other properties of solutions; indeed, this is the problem that eluded [Wilhelm] Pfeffer and [Rudolf] Clausius and that was solved by van 't Hoff.[3] From the perspective afforded by these considerations, it is clear that Poincaré's emphasis on relations is not the expression of any particular philosophical position regarding our knowledge of the properties of bodies.

We have established that there were two sources for Poincaré's emphasis on relations: (1) physical theories are often *indifferent* to the nature of the constitution of the things whose behavior they seek to describe, and (2) physics thrives on the discovery of relations which yield various independent and concordant determinations of physical parameters. We must now consider how Poincaré's emphasis on the importance of physics' discovery of such true relations led him to modify his view of atomistic concepts like those involved in the molecular hypothesis.

In "Hypotheses in Physics,"[4] Poincaré took his analysis of the value of atomistic theories as possibly suggestive guides for future research a step further and argued that the question of the atomic constitution of matter concerns "an indifferent hypothesis," meaning by this that it is a hypothesis whose assumption is at best a heuristic aid which complements the cognitive style with which some theorists approach their calculations.[5] But in his (1912a) essay, Poincaré came to recognize that, in assigning this methodological status to the atomic hypothesis, he had conflated the question of whether we might dispense once and for all with continuity with the more restricted question of whether the molecular and atomic hypotheses can ever achieve the status of scientific fact.[6] By its very formulation, the former question seems to invite a picture of unstable vacillation between alternative resolutions.[7] But Poincaré argued that this is *not* the situation with the latter question in light of the justification the molecular hypothesis acquired with the recent discovery of appropriate relations. The principal one among these relations is the identity of the mean kinetic energies of the Brownian particles and the molecules comprising the fluid in which they are suspended. It afforded a means of empirically determining the values of molecular parameters that had previously been lacking. And as we learned from our discussion of Perrin, Thomson, and others, the concordance between this determination of Avogadro's constant and the determinations of it that are afforded by its relation to parameters of a wholly different character was a key premise in the argument that molecular reality no longer rests on merely hypothetical reasoning.

The *robustness* of the empirical determination of N on which Poincaré laid such stress required the discovery of its functional relation to a variety of empirically determinable parameters, all of which are concordant in the values they yield for N. These results were compelling precisely because of the independence of the sources of evidence they employed. As Poincaré saw, this contrasts with the case where the novelty of a phenomenon is only apparent, as it is when its connection with those phenomena for which the hypothesis was originally adduced is so close that any hypothesis which accounts for the one, "must by this very fact account for the [other. . . .] This is not so when experience reveals a coincidence which could have been anticipated and could not be due to chance, and particularly when a numerical coincidence is involved. Now, coincidences of this type have, in recent times, confirmed the atomistic concepts."[8] Such coincidences are the true relations which, Poincaré argued, constitute the legacy of modern physical theories.

Like others before him, Poincaré was adamant in questioning the persuasiveness of the hypothetical reasoning that is characteristic of the method of hypothesis: Such reasoning can yield only an assessment of the molecular hypothesis as an indifferent hypothesis. What Poincaré argues cannot be assigned to chance is the robustness of the empirical determination of N and of those molecular parameters which are functionally dependent on it. The isolation of the relations that established the well-foundedness of the properties of molecules—the isolation of the relations which facilitated the determination of their properties by robust, theory-mediated measurements—revealed a surprising connection between Poincaré's views by showing how relations involving a variety of independent parameters bear on the accessibility of molecular reality. It changed Poincaré's evaluation of the molecular hypothesis and led him to elevate the hypothesis from the status of an indifferent hypothesis to the level of scientific fact:

> Since we have a second means to count molecules, absolutely independent from that of M. Perrin, let us compare them; this time we find 650 thousand billion billion. This is a surprising agreement, quite unexpected. You can well understand that a few thousand billion billion doesn't make a difference.
>
> This time, there is cause for wonder, especially since more than a dozen entirely independent processes that I would not be able to enumerate without tiring you lead us to the same result. If there were more or fewer molecules per gram, the brightness of the blue sky would be entirely different; incandescent bodies would radiate more or radiate less, and so on. (Poincaré 1912a, p. 224)

There then follows Poincaré's often-quoted remark, to which we called attention in Section 2.4, that there is no denying that we see molecules.

As we saw in our discussion of Perrin, well-foundedness is a desideratum that the application of hypothetico-deductive reasoning and the method of inference to the best explanation simply overlook. And although the satisfaction of this desideratum does not suffice to establish the molecular-kinetic *theory,* Poincaré argued that it provided an adequate basis for securing molecular *reality*. This recognition did not rest on a change of view about proper scientific methodology—still less on a conversion from one philosophical persuasion to another—but is wholly explained by Poincaré's consistent application of his ideas about the limitations of the method of hypothesis and the need for theoretical parameters to be empirically based in true relations.

3.2   Robustness versus Consilience

It is important to distinguish a justification of the molecular hypothesis which, like the one presented here, appeals to the robustness of theory-mediated measurements of molecular parameters, from one which appeals to Whewell's consilience of inductions.[9] The manner in which we are able to obtain information about molecular parameters like N—and the way in which the information so obtained is held to bear on the truth of the molecular hypothesis—is very different in these two accounts. A justification in terms of consilience begins from the premise that when the molecular hypothesis enters into a variety of theoretically different calculations of the same predicted value of N, its confirmation is different in kind from the confirmation it receives from its occurrence in a single calculation. In this respect, consilience constitutes an *enhancement* of the method of hypothesis, rather than a challenge to it.

It is not my purpose to reject consilience-based arguments in favor of the molecular hypothesis, or to argue that consilience played no role in the eventual acceptance of the molecular hypothesis by Poincaré and other of its critics.[10] But it is not the decisive justification of the molecular hypothesis that, I have been urging, was suggested by the discovery of robust, theory-mediated measurements. Moreover, this justification has certain strengths which are not shared by alternatives which appeal to consilience.

An account of Whewell's views on consilience that raises exactly the issues I wish to highlight is given by Laudan:

> If, instead of being able to predict only phenomena of the same kind as the hypothesis was invented to explain, we can explain and predict with its help, cases *of a different kind* (relative of course, to other theories), then we have *indubitable* evidence for the truth of our theory:
>
> > These instances in which this [consilience] has occurred, indeed, impress us with a conviction that the truth of our hypothesis is certain. No accident could give rise to such an extraordinary coincidence. No false supposition could, after being adjusted to one class of phenomena, exactly represent a different class, where the agreement was unforeseen and uncontemplated. That rules springing from remote and unconnected quarters should thus leap, to the same point, can only arise from *that* being the point where truth resides.[11]

Laudan then comments that in the conclusion of this passage Whewell

> seems to be suggesting that this attitude is *logically* justified, that it is simply impossible in principle that any hypothesis could achieve a consilience unless it were the true

hypothesis for explaining the phenomena under investigation. But neither in this passage nor elsewhere does Whewell offer any valid argument to support his logical (as opposed to his psychological) claim.[12]

In support of his assessment, Laudan presses the point that it is not possible to transform an argument based on the predictive success of a hypothesis—even predictive success in the form of consilience—into a logical demonstration of its truth. This undoubtedly correct observation would have to be taken into account in any assessment of the utility of a consilience-based justification for an existence claim. But the cogency of the justification existential hypotheses receive from robust theory-mediated measurements does not rest on the possibility of turning a nondemonstrative inference into a demonstrative one. The fact that the parameters that qualify the hypothetical entities are empirically determinable sets them apart from entities whose properties are accessible to us only on the basis of the predictive success of the explanatory hypotheses in which they occur. Consilience may be a significant refinement of the notion of predictive success that is appealed to by scientific realists in support of their view that successful prediction would be miraculous if the theory were not true. But this just highlights the difference between the goals of scientific realism—and the method of argument by which it seeks to achieve those goals—and our use of robust theory-mediated measurement to establish existential hypotheses.

The fact that we can empirically determine the properties of molecules by robust theory-mediated measurements shows them to be epistemically accessible in a way that hypothetical entities which are merely explanatorily successful are not. As we argued earlier, this is precisely what Poincaré should be understood as having conveyed with his rhetorical remark that in light of Perrin's discoveries, there is no denying that we see molecules. Consilience of inductions, like hypothetico-deductive reasoning and inference to the best explanation, misses the distinctive contribution of theory-mediated measurement to the justification of the molecular hypothesis: Molecular reality is secure because such measurements show molecules to be epistemically accessible rather than merely indispensable to the explanatory hypotheses in which they occur.

3.3   Poincaré and Scientific Realism

The representation of Poincaré as an early advocate of a scientific realist defense of science rests largely on an interpretation of certain of his appeals to chance. One appeal which has been cited in support of this interpretation occurs in "Hypotheses in Physics":

> We have verified a simple law in a considerable number of particular cases. We refuse to admit that this coincidence, so often repeated, is a result of mere chance and we conclude that the law must be true in the general case.
>
> Kepler remarks that the positions of the planets observed by Tycho are all on the same ellipse. Not for one moment does he think that, by a singular freak of chance, Tycho had never looked at the heavens except at the very moment when the path of the planet happened to cut that ellipse. . . . [I]f a simple law has been observed in several particular cases, we may legitimately suppose that it will be true in analogous cases. To refuse to admit this would be to attribute an inadmissible role to chance.[13]

According to the scientific realist interpretation, this passage shows Poincaré to be endorsing the notion that the best explanation of the predictive success of a theory like Kepler's is that it is approximately true, since its success would otherwise be a miracle—"a singular freak of chance." In light of our reconstruction of the nature of Poincaré's argument for the molecular hypothesis as a justification that allows for reservations about the truth of the theories of which it is a part, there should be a far less tendentious and more compelling interpretation of the passage. And indeed there is: Poincaré's remark is entirely captured by the claim that, on the assumption that Tycho's sampling of planetary positions is fair, *ordinary inductive reasoning* suffices to account for the methodological basis of what is nondemonstrative in Kepler's derivation of his law.

Given Poincaré's consistent hostility to the method of hypothesis, the scientific realist suggestion—that for Poincaré, the truth of a hypothesis can be inferred from its predictive success—is even less plausible in cases like the molecular hypothesis, and similar constructive components of theories, than it is in the case of Kepler's law. Scientific *structural* realists have attempted to address this point by distinguishing an inference to *structure* from one to *content* or *ontology*, and then characterizing the realism they advocate as a combination of realism about structure and agnosticism about ontology. Poincaré was certainly deeply skeptical of the possibility of forming a scientific judgment about atomism, which is of course an ontological and contentual assumption even when it is glossed as the assertion that matter has a granular structure. So his eventual change of view—focused as it was on an ontological or contentual question about the constitution of matter—should be deeply puzzling on a structural realist interpretation of him. And it counts against both structural and nonstructural scientific realist interpretations that Poincaré's rejection of his earlier skepticism did not rest on his having come to embrace the thesis that the predictive success of the molecular hypothesis would be miraculous were it not true.

The basis for Poincaré's mature view of molecular reality was the same as Perrin's: Once it could be shown to be epistemically accessible by robust theory-mediated determinations of the properties of molecules, molecular reality was on the path to being secured. But this is a justification of the molecular hypothesis that can be maintained in the face of serious questions about the truth of our theories of molecular reality. It would therefore be a mistake to represent Poincaré as having mounted a defense of science that was a pre-echo of scientific realism. Poincaré's view is more nuanced than this and combines both realist and nonrealist aspects. It is realist insofar as it fully supports the reality of entities that transcend observation; but it does so on the basis of experimental and theoretical advances within science that are related to measurement. By *not* resting on the supposition that the predictive success of theories would be miraculous were they not at least approximately true, Poincaré's position differs from standard conceptions of scientific realism. It is, however, important to emphasize that his methodological remarks avoid certain aspects of realism without falling victim to the then-emerging positivist consensus—represented perhaps most prominently by Mach—that theory should be reduced to observation. Mach expressed this thesis with a distinction between theories that use only "direct descriptions"—by which he meant abstract principles which employ only descriptions of what is observable—and theories that incorporate "indirect descriptions," which go beyond what is observable. It is, according to Mach,

> not only advisable, but even necessary, with all due recognition of the helpfulness of theoretic ideas in research, yet

gradually, as the new facts grow familiar, to substitute for indirect descriptions direct description, which contains nothing that is unessential and restricts itself absolutely to the abstract apprehension of the facts.[14]

Poincaré nowhere commits himself to Mach's view of the preferred relation of theory to observation or to what is typically represented as Mach's view of the prominence of observation. And to the extent that for Poincaré, the value of theories is "instrumental," this must include their instrumental value in revealing relations which guide us to a representation of the constitution of a reality that lies hidden from observation.

3.4 Russell and Poincaré

In his review of *Science and Hypothesis,* Russell said of Poincaré that he holds "[q]uestions concerning the real, as opposed to the relation of real things, . . . to be illusory and devoid of meaning (pp. xxiv, 163)." He then continues

Certainly we have much more belief in the accuracy of our perceptions of relations than in that of our perceptions of qualities. When we see green in one place and red in another, we are willing to believe that secondary qualities are subjective, but not that the fact of difference between what is in the two places is an illusion. It is only by holding fast to relations as perceived that science manages, on an empirical basis, to construct a world so different from that of perception. Why we should trust in our perception of relations I do not know; but it is a fact that we do so. But I do not see how it can be maintained that questions as to the qualities of real things are *unmeaning*. The proposition amounts to this, that if *a* really exists, a statement about *a* has no *meaning* unless it asserts a relation to a *b* which also really exists. The fact seems to be, not that such propositions are unmeaning, but that, except in psychology, they are unknowable. We may even push the theory further, and say that in general even the relations are for the most part unknown, and what is known are properties of the relations, such as are dealt with by mathematics. And this, I think, expresses substantially the same view as that which M. Poincaré really holds.[15]

The passages Russell cites show Poincaré to be at the very least equivocal between the unknowability of certain questions and their meaninglessness. In the first passage (p. xxiv), Poincaré writes that "the aim of science is not things in themselves, as the dogmatists in their simplicity imagine, but the relations between things; outside those relations there is no reality knowable." While in the second (p. 163) he defends the exclusion of the question of the truth of "the images we have formed to ourselves of reality," on the ground that such "questions which we forbid you to investigate, and which you so regret, are not only insoluble, they are illusory and devoid of meaning." Both passages are part of a broader polemic against certain "dogmatists."[16]

The view Russell attributes to Poincaré is evidently very close to one he himself favors. Russell's own theory arises from reflection on what can be inferred about the causal sources of the objects of perception on the basis of our perceptual experience. Coming as he does from the tradition of British empiricism, Russell is skeptical about the thesis that we can know of the qualities of material objects that they are the same as the properties that qualify our experience of them. The passage we have quoted from his review considers whether this doubt should not also apply to the relations between material objects. The relation Russell chooses, in order to illustrate that such doubts are not always warranted, is the relation that holds between two "places" when what qualifies something in one of the places is *different* from what qualifies something else in the other. Notice that this is a purely logical relation, and that since, for Russell, mathematics is just a development of logic, this concession does not contradict what Russell proposes as the central tenet of his theory, namely, that what can be known of the material world is only its mathematical properties. But Russell's main point in this passage is to argue for an emendation to what he takes to be Poincaré's view that relations between material objects are unlike their qualities in being knowable. Russell proposes that Poincaré "push [his] theory further, and say that in general even the relations are for the most part unknown, and what is known are properties of the relations, such as are dealt with by mathematics."[17] This is Russell's structuralist reconstruction of Poincaré.

Russell's assimilation of Poincaré's views to his own version of structuralism receives considerable textual support from Poincaré's paper, "Science and Reality." The paper occurs as the final chapter of *The Value of Science* (Poincaré 1905), which was published in the same year as Russell's review. In this paper, Poincaré appeals to the example of the relations of sameness and difference to illustrate the possibility of making objective assertions about phenomena that are themselves irreducibly subjective. The point of Poincaré's remarks is not to draw a contrast between the attribution of relations and the attribution of qualities to physical objects, but to exhibit the possibility of an objective comparison of a relation that holds between my sensations with the relation that holds between the sensations of another, however subjective the sensations themselves. Here is the relevant passage:

> The sensations of others will be for us a world eternally closed. We have no means of verifying that the sensation I call red is the same as that which my neighbor calls red.
>
> Suppose that a cherry and a red poppy produce on me the sensation *A* and on him the sensation *B* and that, on the contrary, a leaf produces on me the sensation *B* and on him the sensation *A*. It is clear we shall never know anything about it; since I shall call red the sensation *A* and green the sensation *B,* while he will call the first green and the second red. In compensation, what we shall be able to ascertain is that, for him as for me, the cherry and the red poppy produce the *same* sensation, since he gives the same name to the sensations he feels and I do the same.
>
> Sensations are therefore intransmissible, or rather all that is pure quality in them is intransmissible and forever impenetrable. But it is not the same with relations between these sensations. (Poincaré 1905, authorized 1913 Halsted translation, p. 348)

It is therefore not surprising that in a Letter to the Editor of *Mind,* published a year after Russell's review, Poincaré should have concurred with Russell's understanding of him:

> I have said that questions related to the qualities of real things are *unmeaning* because for a question to make sense, we need to be able to conceive of an answer that would make sense. Now this answer could only be made

with words and these words would only be able to express psychological states, i.e. subjective secondary qualities, that would not be those of real things. At the end of the paragraph he devotes to this question, Mr. Russell says: "We may even push the theory further, and say that in general even the relations are for the most part unknown, and what is known are properties of the relations, such as are dealt with by mathematics. And this, I think, expresses substantially the same view as that which M. Poincaré really holds." Russell is not deceiving himself. That is my thought.[18]

It should, however, be abundantly clear from our earlier discussion that Poincaré's views regarding the qualities of experience—even when expanded to include the qualitative dimension of relations as we experience them—are entirely orthogonal to his views on true relations. Those views represent a contribution to our understanding of the character of contemporary developments in physics—most significantly, to our appreciation of the epistemic status of the molecular hypothesis—rather than a general theory of our knowledge of the external world in the tradition of Russell's causal theory of perception. This point has been missed in the current tendency to see in Poincaré's remarks on true relations an anticipation of Russell's structuralism, or of structural realism.

4 Quantum Reality

4.1 Bohr on the Primacy of Classical Concepts

By the "primacy of classical concepts" for our understanding of quantum mechanics I mean—and I take Bohr to have meant—their primacy in the description of experimental results pertinent to the development and confirmation of the theory. This is clear from his earliest writings on the "new" quantum theory. It is, for example, completely explicit in his 1927 essay, "The Quantum Postulate and the Recent Development of Atomic Theory":

> The quantum theory is characterized by the acknowledgment of a fundamental limitation in classical physical ideas when applied to atomic phenomena. The situation thus created is of a peculiar nature, since our interpretation of the experimental material rests essentially upon classical concepts.[1]

Bohr's thesis is not about the primacy of classical concepts in the *theoretical* claims of any system of future physics. Their primacy is *evidentiary*.

There is a general point about evidentiary frameworks that emerged from our analysis of the discoveries of Perrin and Thomson involving molecular and subatomic reality that bears on the initial or prima facie plausibility of the primacy of classical concepts in Bohr's sense: A framework within which experiments are designed and their results reported and assessed must contain standards that enable agreement regarding the cogency and intended significance of experimental results. But this requirement can be satisfied only if the concepts of the evidentiary framework have clear and generally agreed-upon criteria of application and only if the principles which employ them are reliable and of sufficient "validity." The "validity" the principles enjoy may well be limited and not universal; nor need it reflect the truth of the presuppositions of the framework within which the principles are formulated. This notion of limited validity may be illustrated by a case that is familiar from our earlier discussion.

Stokes's Law of Fall expresses the relation of the rate of fall of a spherical object to its density and the density and viscosity of the medium in which it is immersed. The law guided Perrin's and Thomson's determinations of the properties of the molecular and subatomic constituents of matter, even though it was thought unlikely that the relation it expresses could be expected to hold for spherical objects of the dimensions required by their applications of it. Nevertheless, Stokes's law isolates what, in Poincaré's terminology is a "true" relation—within a limited domain—between the rate of fall of spherical objects, density, and viscosity that is preserved under a change of application from the continuous media for which it was initially devised to discrete media. Despite its limitations, it served as a reliable guide to the determination of the molecular diameter and the charge on an electron. It illustrates the fact that the presuppositions of the principles which underlie an evidentiary framework might be false—and even *known* to be false—and the principles themselves of only limited validity, without losing their effectiveness for probing the evidence for a theoretical claim, or refining the determination of a theoretical parameter. This process is an iterative one involving an intricate interplay of experimental design and successive corrections of those "true relations" that are essential to securing more refined approximations to the values of theoretically significant parameters.

This conclusion is perfectly general, although it hardly establishes Bohr's specific insistence on the primacy of classical concepts, it bears on its initial plausibility when the thesis is understood as a claim about the role of classical concepts in the *evidentiary framework* of quantum mechanics. So understood it is clear that Bohr's thesis is altogether different from the idea that it is tied to a dogmatically conservative view regarding the necessity—perhaps even the a priori necessity—of classical physics. Understood as a thesis about the epistemic framework within which physical theories are evaluated, the thesis of the primacy of classical concepts is entirely compatible with the idea that the principles and presuppositions of the classical framework are radically mistaken and incapable of providing an adequate theoretical basis for physics. Insofar as Bohr's thesis has sometimes been rejected because of the pervasive bias that the concepts framing our methodological practices must be based on a framework of assumptions which are true, it has been rejected for the wrong reasons. Only the limited validity of the principles of classical physics is required for the classical framework to play the evidentiary role Bohr claimed for it.[2] But this is not the only reason for the dissatisfaction which Bohr's thesis of the epistemic primacy of classical concepts has elicited. In a letter of October 13,

1935, Schrödinger stated his difficulty with Bohr's thesis very forcefully:

> You have repeatedly expressed your definite conviction that measurements must be described in terms of classical concepts. For example, on p. 61 (94–95) of the volume published by Springer in 1931: "It lies in the nature of physical observation, nevertheless, that all experience must ultimately be expressed in terms of classical concepts, neglecting the quantum of action." And ibid p. 74 (114): "the invocation of classical ideas, necessitated by the very nature of measurement." And once again you talk about "the indispensable use of classical concepts in the interpretation of all proper measurements." True enough, shortly thereafter you say: "The removal of any incompleteness in the present methods of atomic physics . . . might indeed only be effected by a still more radical departure from the method of description of classical physics, involving the consideration of the atomic constitution of all measuring instruments, which it has hitherto been possible to disregard in quantum mechanics."[3]
>
> This [last passage] might sound as if what was earlier characterized as inherent in the very nature of any physical observation as an "indispensable necessity," would on the other hand after all just be a, fortunately still permissible, convenient way of conveying information, a way we presumably sometime will be forced to give up. If this were your opinion, then I would gladly agree. . . .
>
> [But] I think that the fact that we have not adapted our thinking and our means of expression to the new theory cannot possibly be the reason for the conviction that experiments must always be described in the classical

manner, thus neglecting the essential characteristics of the new theory. It may be a childish example [; I use it] only to say briefly what I mean: after the elastic light theory was replaced by the electromagnetic one, one did not say that the experimental findings should be expressed—just as before—in terms of the elasticity and density of the ether, of displacements, states of deformation, velocities and angular velocities of the ether particles.

It is clear that Schrödinger readily concedes the practical utility of the appeal to classical concepts in the characterization of measurements relevant to the quantum theory. His question concerns the necessity of such an appeal and the basis for Bohr's belief that the invocation of classical ideas is required by the very nature of measurement.

Bohr's response to Schrödinger is contained in his letter of October 26, 1935 (Bohr 1935b / 1996, pp. 511–512; all italics Bohr's):

My emphasis of the point that the classical description of experiments is unavoidable amounts merely to *the seemingly obvious fact that the description of any measuring arrangement must, in an essential manner, involve the arrangement of the instruments in space and their functioning in time, if we shall be able to state anything at all about the phenomena.* The argument here is of course first and foremost that in order to serve as measuring instruments, they cannot be included in the realm of application proper to quantum mechanics.

This reply reiterates the claim that only a classical description of a measurement arrangement is capable of capturing its epistemological function in an experiment without fully explaining why this *must* be the case. Insofar as Bohr can be understood to have addressed Schrödinger, his answer must be bound up with the idea that a quantum-mechanical description, unlike a classical one, leaves out the arrangement of the instruments in space *and* their functioning in time. Bohr appears to be claiming that this is something any description of measuring instruments must include in order to play the epistemic role they do. Beginning from this perspective, I take Bohr's allusion to the functioning of the instruments in time and their arrangement in space to be a point about the *completeness* that attaches to a classical description by virtue of its inclusion of *both* spatial arrangement and dynamical behavior. The kind of completeness that attaches to classical descriptions is given up in the transition from classical to quantum mechanics. A classically complete description of the systems under investigation by the new theory is replaced by a form of description which, though incomplete relative to classical expectations, exhibits "complementarity." But for reasons of epistemic accessibility, the descriptions of the instruments with which we probe these systems and evaluate the theory's claims about them must retain their classically complete character.

Rather than attempt a detailed interpretation of Bohr's difficult and complex view, I intend to develop an analysis of classicality that, although not explicit in Bohr's writings, supports a thesis of the epistemic primacy of classical concepts for our understanding of quantum mechanics. In the course of presenting this analysis, I will sometimes compare it with explicit formulations of Bohr's to support the claim that the analysis is at least "Bohrian" in spirit. But I do not advance this analysis as a piece of Bohr interpretation, and I do not claim Bohr's authority on its behalf.

On the explication of *classicality* that I believe is relevant to our understanding of quantum mechanics, the central characteristic of a framework or theory whose concepts are classical is the commutativity of the algebra of physical concepts—the parameters, physical magnitudes, and dynamical variables—with which it characterizes physical systems. Equivalently, classicality consists in the Boolean character of the algebra of all the properties or propositions that are associated with each physical system. On this view, classicality is a characteristic that attaches to the *interrelations* of the physical concepts of a theory, rather than to the concepts themselves. As a consequence, Schrödinger's attempt to undermine the thesis of the primacy of classical concepts by appealing to the possibility of introducing into physics new and hitherto unfamiliar concepts misidentifies the source of the difference between the classical and quantum-mechanical frameworks and what is essential to the claim that classical concepts retain their *epistemic* primacy—their primacy in the evidentiary framework—when we pass from classical to quantum mechanics. On my reconstruction, the fundamental difference between classical and quantum mechanics, so far as the thesis of the primacy of classical concepts is concerned, does not lie in the concepts themselves but in how they are interrelated by the functional relations that hold among them.

Although this idea of classicality is not one that Bohr explicitly articulated, the noncommutativity of the "quantum formalism" is something to which he frequently alludes when he remarks on the role of the quantum of action in the commutation relations between classically conjugate parameters. Complementarity is clearly central to his view of the theory, and Bohr invokes it to address the significance of the fact that of two conjugate parameters which together comprise a classical representation of a system, quantum mechanics allows us to probe only one of them at a time. Significantly, the theory imposes no restriction on which parameter we might probe, and it allows complete freedom in our choice of which of two complementary descriptions to apply. Thus the point of complementarity may well have been to address the noncommutativity of the algebra of physical concepts—the so-called *observables* that quantum mechanics associates with a physical system. In any case, the emphasis on noncommutativity is a feature of my explication of classicality that is not without parallel in Bohr. But Bohr was also concerned to show that despite the limitation noncommutativity imposes on what we can experimentally discover about a quantum-mechanical system—despite the fact that it leads to a representation that is incomplete *relative to classical mechanical expectations*—the quantum-mechanical representation is, in a more basic and important sense, *complete*. As I indicated earlier, the completeness of classical descriptions, by contrast with the complementarity of the quantum-mechanical case, informs his reply to Schrödinger on the epistemic primacy of classical concepts and their unique position in the evidentiary framework of the quantum theory. The nature of the completeness that attaches to quantum-mechanical descriptions and the controversy surrounding it is a topic to which we will return.

The thesis of the primacy of classical concepts is naturally paired with the idea that measurements probe quantum systems in order to elicit their effects on systems which are conceptualized as classical. Notice that I do not say "*are* classical." If quantum mechanics is true, then *all* physical systems are quantum mechanical. However this is compatible with holding that the evidentiary framework of quantum mechanics is classical. We extrapolate from such effects to the values of the observables of the measured system, and we apply the mathematical framework of quantum mechanics to guide our conditional expectation of the likelihood of occurrence of other possible effects, should we undertake to determine another of the system's observables. The totality of such possible effects has a structure which is determined by the system of functional relations among the observables of a quantum system. This structure is represented by the noncommutative algebra of Hermitian operators acting on a Hilbert space, an algebra which is not embeddable in a commutative algebra, and for which there is no possible truth-value assignment to the subalgebra of properties or propositions associated with the operators of the algebra. The absence of such a truth-value assignment is completely alien to the conceptual framework of classical physics, and to the evidentiary framework with which the quantum-mechanical effects elicited by measurement interactions are recorded, and their methodological significance assessed.

The shift in the algebraic structure of observables and properties which marks the transition from classical to quantum mechanics is a radical departure, even by the standard set by the transition from Newtonian ideas that characterized the special and general theories of relativity. In the case of special relativity, it is possible to define within Minkowski space-time a unique relative simultaneity relation—simultaneity relative to an inertial frame. Taken by itself, each such frame is "classical" in the sense that its hyperplanes of simultaneous events are related as they are in Newtonian space-time, with spatial and temporal quantities definable relative to an inertial frame as they are in the prerelativistic or Newtonian case. Each of the space-times of this totality is itself a *classically complete* collection of kinematic possibilities. But Minkowski space-time is distinguished from Newtonian space-time by the fact that the *totality* of such possible "Newtonian" frames comprises a non-Newtonian spatiotemporal structure. The discovery of this structure is arguably the main conceptual innovation of special relativity.

Bohr expressed something close to this view of the nature of the conceptual shift introduced by special relativity when he wrote that

> [a]lthough in [special relativity] use is made of mathematical abstractions such as a four-dimensional non-Euclidean metric, the physical interpretation for each observer rests on the usual separation between space and time, and maintains the deterministic character of the description.[4]

The important point of similarity is the emphasis Bohr gives to "the usual separation between space and time," which he attributes to each "observer," where an observer is evidently intended to be interchangeable with an inertial frame of reference. Such a separation is of course characteristic of pre-relativistic conceptions of space and time. So understood, Bohr is calling attention to the fact that in the special relativistic case, it is possible to model the kinematic possibilities as they are represented by classical physics.

In the case of quantum mechanics, each observable is represented by a Boolean algebra of possible properties corresponding to the possible values of the observable, and this is reflected in the Boolean algebra of possible effects that are elicited by measurement interactions involving the determination of the value of the observable. The totality of all possible observables of a quantum system, as well as that of their associated measurement effects, comprises a nonclassical structure whose discovery constitutes the principal conceptual novelty of the quantum theory. But by contrast with Minkowski geometry and special relativity, where each space-time associated with an inertial frame is classically complete with respect to the kinematic possibilities it contains, the classical components of the quantum-mechanical representation of the possible properties of a physical system are restricted to those possibilities that are associated with only one of two classically conjugate observables. This yields a description of the properties of a physical system that is incomplete relative to our classical conception of a physical state in a way that contrasts with special relativity and the complete representation of kinematic possibilities associated with each Newtonian frame.

The case of general relativity affords yet another perspective on the uniqueness of the transition from classical to quantum mechanics. In a paper devoted to the exposition and reevaluation of Kant's doctrines in light of developments in twentieth-century physics, Michael Friedman recounts the famous eclipse observations of 1919:

> [T]he space of the observatory is so small relative to the corresponding space of the cosmos (here extending well beyond the solar system) that it is, for all intents and purposes, *infinitesimally* small; so its geometry, even according to the general theory of relativity, remains Euclidean. In this sense, it is still reasonable to view the use of this geometry as an a priori constitutive presupposition of the empirical observations that serve as tests for the theory, even though the global (cosmic) space employed by the theory is measurably non-Euclidean.[5]

Friedman concludes from this that "a generalized and extended version of the Kantian conception of causal necessity, involving both its conceptual and intuitive components, remains a viable option." If we replace Friedman's allusion to causal necessity with the claim that the Kantian conception, in the form of the assumption of a flat or Euclidean spatial geometry, remains a viable option for the *evidentiary framework* of relativity, then there is an interesting connection with how I am suggesting we should understand Bohr's thesis of the primacy of classical concepts for the evidentiary framework of quantum mechanics. The assumption of flatness in the "infinitesimally small" is generally conceded to be viable even if the space of a region of the dimensions of Sobral observatory is *non*-Euclidean "in reality," Euclidean only for all practical purposes. Of course, what is viable is not the claim that space *is* flat or Euclidean, but the claim that the intuitive and conceptual components of our *evidentiary framework* are, as in the Kantian conception, Euclidean. Epistemic "necessity" can attach to our current intuitive and conceptual components, in this sense, even when they fail to capture the true geometrical structure of space.

There is an important and not widely known fact about general relativity, Newtonian gravitation theory, and the geometry of space, that was independently discovered by David Malament and Jürgen Ehlers.[6] In its Trautman-Cartan formulation, Newtonian gravitation is recoverable as a limit of general relativity, as *c* goes to infinity, only if in this limit, the geometrical structure of space is Euclidean. The epistemic interpretation of Friedman's point does not depend on this fact. It is, however, of some interest for Friedman's formulation as a point about Kant and causal necessity, since it shows that the idea that space is Euclidean—even the idea that it is in a relative sense causally necessary that it is Euclidean—is supported by the Malament-Ehlers theorem that the space of Newtonian gravitation theory is Euclidean in an appropriate classical limit of the *general-relativistic* theory of gravitation.

By contrast, the thesis of the primacy of classical concepts in the case of quantum mechanics is a purely epistemic claim about the Boolean character of the evidentiary framework of the theory. The relation of quantum mechanics to the Boolean structure of the possible states of a physical system in classical mechanics might be radically different from the relation of the general-relativistic theory of gravitation to the geometry of space in Newtonian gravitation without calling into question the epistemic primacy of the classical framework. Both the Euclidean structure of space and the Boolean structure of the possible states of a physical system are pervasive features of our evidentiary framework. But even if it is not recoverable as an appropriate limit of quantum mechanics, the evidentiary framework of quantum mechanics would nevertheless retain its Boolean character. Let us consider another, more familiar, point of comparison between the transition from classical mechanics to general relativity and the transition from classical mechanics to quantum mechanics.

In general relativity if we focus on *spatiotemporal* structure, rather than spatial structure, we find that each tangent space is a complete representation of the local geometry of space-time. But while each of the tangent spaces is flat or "classical," the totality of all the tangent spaces is not flat. Now one might argue that quantum mechanics displays an analogous feature: The algebra of quantum-mechanical propositions is representable as a family of "locally" classical (i.e. Boolean) algebras, but the algebra *as a whole* is not Boolean. Yet this comparison can be seriously misleading if one overlooks an important difference between the quantum-mechanical and the relativistic cases: None of the Boolean subalgebras of the non-Boolean structure is a *classically complete algebra of possible propositions* in which the states of physical systems as we experience them can be located. That is to say, in the quantum-mechanical case, none of the "locally" classical algebras has the completeness vis á vis classical possibilities for the combination of properties or propositions that is exhibited by the completeness of the local spatiotemporal geometry of the tangent spaces of the spatiotemporal manifold of general relativity.

An insistence on the primacy of the classical level for the description of the evidentiary framework of quantum mechanics involves considerations that are fundamentally different from what occurs in the curved space-times of general relativity. In the latter case, we can appeal to the idealization that is possible in "infinitely small" regions, and we can even support this idealization for the Euclidean geometry of *space* within general relativity by appealing to the Ehlers-Malament theorem relating the gravitation theory of general relativity to Euclidean space and the Trautman-Cartan formulation of Newtonian gravitation. But the representation of the algebra of quantum-mechanical propositions by a collection of Boolean algebras does not translate into a representation of this algebra as a collection of classical possibilities in the way in which the relativistic conception of the global geometry of space-time can be conceptualized as a totality of locally flat space-time. For the analogy to succeed, it isn't enough that the subalgebras should be *Boolean,* they must also be classically *complete* structures of possibilities among properties in the way that the local space-times are classically complete structures of spatiotemporal possibilities. I believe it is the failure of the analogy at just this point that marks the uniqueness of the conceptual shift exhibited by the quantum-mechanical transition from classical ideas.

We can express the situation in Bohr's terms as follows: Each Newtonian frame is "visualizable" as a totality of kinematic possibilities. But in the case of quantum mechanics, where the structure of possible properties is so restricted that it allows the application of only one or another of two classically conjugate parameters, we are not in a position to view the space of possible properties as the properties of a single *visualizable* entity. We can visualize the effects quantum systems have on the instruments which probe them, provided these instruments are conceptualized within the evidentiary framework of classical mechanics. The classical framework is not recoverable as an appropriate limit from the framework of quantum mechanics, but this fact does not compromise its methodological role as the evidentiary framework of quantum mechanics. The radical disparity between the algebraic structure of the classical and quantum-mechanical frameworks is not a problem that must be overcome, but is rather the true basis for the uniqueness of quantum mechanics in the evolution of physical theories that Bohr sought to highlight by his insistence on the methodological primacy of classical concepts.

I am aware of two recent discussions of Bohr that are similarly sympathetic to the idea that experiments must be described using classical concepts. But by contrast with the view presented here, these approaches impose as a condition of adequacy the requirement that the classical framework should be recoverable as an appropriate approximation from quantum mechanics. Let us briefly consider these two discussions.

(1) Camilleri and Schlosshauer (2015, pp. 73–83) emphasize the distinction between the mandatory use of classical *concepts* as opposed to classical *theories*. They note that

[w]hile Bohr often left it to his readers to decipher the precise meaning of ambiguous phrases such as "classical description," in his more deliberate moments he did take care to distinguish between the use of classical concepts (such as position and momentum) and classical dynamical theories. In his reply to the EPR [Einstein-Podolsky-Rosen] paper, for example, Bohr emphasized the necessity of using "classical concepts in the interpretation of all proper measurements, even though the classical theories do not suffice in accounting for the new types of regularities with which we are concerned in atomic physics."[7]

Camilleri and Schlosshauer use the distinction between theories and concepts to observe that we are not committed to the approximate *truth* of classical physics by the claim that experiments must be described using classical concepts. This is a point on which we have also insisted. However, Camilleri and Schlosshauer go on to argue that in order to ensure the coherence of his emphasis on classical concepts, Bohr must show that classical mechanics is recoverable from quantum mechanics as an *approximation,* since, without such a demonstration, classical concepts cannot serve the role Bohr assigned them in measurement:

> Bohr's epistemological explanation for why we must use a classical description . . . begs the question of what dynamical features of a macroscopic system entitle us to neglect the "quantum effects." Bohr . . . appears to simply assume that there exists a macroscopic "region where the quantum-mechanical description of the process concerned is effectively equivalent with the classical description" (Bohr, 1935b, p. 701). Thus we are led to ask: How is it that clas-

sical physics can be employed, at least to a very good approximation, under certain dynamical conditions (typically those corresponding to measuring scenarios)? This is a salient question, given that, strictly speaking, the world, as Bohr recognized, is nonclassical. (Camilleri and Schlosshauer 2015, p. 79)

(2) Klaas (N. P.) Landsman (2008, pp. 173–190) develops a related, but more abstract point which is focused on the recovery of the algebraic structure of the observables of classical mechanics in the macroscopic limit:

Bohr time and again stresses that measurement devices must be described classically 'if these are to serve their purpose.' We take this to mean that, although such devices are *ontologically quantum-mechanical* by nature, they become a tool (in fact, the only tool) for the description of quantum phenomena as soon as they are epistemically treated *as if they were classical.* Thus the so-called Heisenberg cut, i.e., the borderline between the part of the world that is described classically and the part that is described quantum-mechanically, is epistemic or (inter)subjective in nature and hence movable. . . . As always, the mathematical implementation of Bohr's philosophical ideas is ambiguous; as far as his doctrine of classical concepts is concerned, we read it as saying that a quantum system described by a noncommutative algebra A of observables is empirically accessible only through commutative algebras associated with A. . . . The simplest kind of commutative algebras associated with A are its (unital) commutative C^* subalgebras; in this paper we need a more subtle limiting procedure to 'extract' a commutative C* algebra of macroscopic observables.[8]

Landsman is certainly right to insist that for Bohr, the Heisenberg cut is epistemic: The demand that the evidentiary framework of quantum mechanics must be expressed "classically" in no way commits us to the privileged status of classical physics in a description of reality. But although classical concepts may be, and presumably are, replete with false presuppositions, this does not compromise the acceptability of their deployment in the framework within which experiments are formulated and experimental results reported and evaluated.

Contrary to Camilleri and Schlosshauer, Bohr's "effective equivalence" of classical and quantum descriptions of the process or measurement commits us *only* to the idea that what we describe as occurring at the level of quantum effects can be transferred to quantum-theoretical predictions about the values of the observables of the systems that are being measured. If this were not possible, the results of measurement could not be understood to bear on the theory's predictions.

It would indeed be a defect if the objects we employ as measuring instruments were not susceptible to a quantum-mechanical representation; Landsman suggests that it would be an acceptable solution to show that this holds in an appropriate limit in which a measuring instrument is represented as infinitely many constituent quantum systems. The difficulty with this proposal is that it fails to establish a relevant connection to the use of the instrument in *measurement*. In any actual measurement the instrument is a quantum system of *finite* complexity; but by hypothesis such systems fail to be classical.

What both Landsman, and Camilleri and Schlosshauer, miss is that even if there are good reasons for seeking an account of classically described systems within quantum mechanics, the ability of such systems to fulfill their function as measuring instruments within the *evidentiary framework* of quantum mechanics neither depends upon nor requires the recovery of their classical description as a quantum-mechanical approximation. By comparison with these proposals, Bohr's thesis of the primacy of classical concepts derives its strength from the clarity with which it confines the use of classical concepts to the evidentiary framework while acknowledging the uniqueness of the relation in which classical and quantum-mechanical forms of representation stand to one another.

4.2   Complementarity, Completeness, and Einstein's Local Realism

It will be useful to have a minimal account of Bohr's notion of complementarity and the position it occupies in his thinking about quantum mechanics, even if the account preserves what appears to be a deliberate ambiguity in Bohr's understanding of the nature of the impossibility of simultaneously determining the values of conjugate parameters: Is the impossibility merely epistemic, the result of a limitation—whatever its source—in our ability to know the values of conjugate parameters? Or does it reflect an indeterminacy in the factual situation we seek to describe? We will attempt to address this ambiguity at a later stage of our discussion. Bohr begins from the assumption that complementarity represents the major conceptual innovation of the quantum theory. But as we noted earlier, it is also key to understanding his conception of the theory's completeness. For Bohr, the theory is incomplete only relative to our classical expectation that it should be possible to simultaneously determine the values of all the parameters that can qualify a physical system. By complementarity, quantum mechanics allows us to probe only one of two classically conjugate parameters at a time, and the classical expectation of a complete description, given by the simultaneous determination of the values of two classically conjugate parameters, is precluded. But by complementarity, it is also the case that the theory imposes no restriction on *which* of the classically conjugate parameters that apply to a quantum-mechanical system we might probe. Because of our freedom to probe *any* parameter, a complementary description exhibits a *kind* of completeness, even though it is not the completeness with which we are familiar from our classical experience. Since for Bohr complementarity is the main conceptual innovation of quantum mechanics, it should be preserved in any account of the theoretical domains with which the theory deals. So for Bohr the kind of completeness which quantum mechanics allows is in some sense the "best possible."

We can distinguish two very different kinds of response to the claim that the completeness quantum mechanics exhibits cannot be improved upon. I will refer to these two responses as the "Bohmian" and "Einsteinian" responses. The distinction between these two responses is "dialectical" in the sense that it is intended to serve an expository purpose in the argument I intend to develop, although it abstracts from important aspects of the historical views suggested by the terms which identify them.[9] The *Bohmian response* interprets the quantum state or wave function *ontically* as a physical property of the system; it proposes to complete quantum mechanics by *supplementing the wave function* with a component which represents a system of possible particle configurations and a dynamics which describes the evolution of this expanded state. This may represent a departure from *Einstein's* view of the quantum state as "statistical." But leaving the interpretive question aside, for the Einsteinian response, the essential point of the comparison of quantum mechanics and its hypothetical completion with the relation between classical statistical mechanics and classical mechanics is the idea that whatever *else* may be true of the states of classical statistical mechanics, they represent our ignorance of the classical phase point of the system. The response seeks to understand the quantum-theoretical statistical description given by the wave function along similar lines.

Since in classical theories such expectations regarding completeness are satisfied as a consequence of the commutativity of their algebras of observables, a natural approach to formulating an Einsteinian completion of quantum mechanics is to recover it from an underlying *commutative* theory. The fact that Einstein was sensitive, from an early date, to the limitations of von Neumann's argument against hidden variables—the same limitations that were later emphasized by Kochen and Specker—shows him to have seriously reflected on the possibility of such a completion of the theory.[10] There are in addition several published remarks which suggest that this is a path he might well have favored.[11] As is now well known, such a development of the Einsteinian response is subject to seemingly insurmountable difficulties in light of certain limitative results—in the form of no hidden variable theorems—of the 1960s. But however slim the prospects of such an Einsteinian extension of quantum mechanics, I hope to show that the exploration of the possibility of such an extension has illuminated the theoretical status of a central principle-theoretic component of the quantum theory itself, one that will play a key role in the interpretation of the theory I will propose. For this reason, the impossibility of such an extension of the theory will be a major focus of our attention.

From the point of view of the discussion which follows, the interest of the Bohmian response consists in its sharp contrast with the Einsteinian response and the clarification of the latter response this contrast affords. The Bohmian response represents an attack on the problem of completeness that is unaffected by those no-hidden-variable theorems that appeal to the algebraic structure of the quantum theory. The theory, to which a Bohmian completion of the quantum state leads, is advanced as an empirically equivalent alternative, one that is not bound by the algebraic constraints that Einsteinian completions of the theory aim to satisfy.[12] This is because the Bohmian response is entirely concentrated on completing the quantum state by pairing the wave function with a system of possible particle trajectories to yield an "ontology" which is otherwise only partially specified by the wave function. As an alternative that is predicated on supplementing the quantum *state,* rather than extending the quantum *theory,* the only condition that a Bohmian alternative recognizes as a reasonable one is that it should recover the quantum theory's correct experimental predictions from an internally consistent theory which is formulated within the framework of an "intelligible" ontology. Unlike the Einsteinian response, neither the success nor failure of the Bohmian program would illuminate the nature of the quantum theory's divergence from classical mechanics over the algebraic structure of the observables with which these theories characterize physical systems.

I am not aware of any writings of Bohr's that explicitly address Bohm's theory. Of Einstein's reaction to Bohm's theory, we know that in a letter to Max Born[13] he dismissed it as "too cheap" a solution, although the sense in which he regarded the solution it offers as too cheap is a matter of speculation.[14] In any case, in so far as Bohr's views on the completeness of quantum mechanics were directed at excluding an alternative which is predicated on the assumption that quantum mechanics is incomplete, it was Einstein's view that concerned him. Bohr would agree with the "naturalness," if not the plausibility, of an approach which sought to satisfy our classical expectations of completeness by recovering the characteristic features of quantum mechanics from an extension of the theory along the lines of the Einsteinian response. But whatever the difficulties, as we now understand them, of implementing a program of the sort advocated by the Einsteinian response, *Einstein's* belief that the quantum-mechanical description of reality is not complete rested on more than "physical intuition." In his paper with Podolsky and Rosen,[15] there is an argument for this conclusion, one which was effectively addressed only some thirty years after its publication.

Following later simplifying developments, it is customary to describe the argument of the EPR paper in terms of directional properties, such as electron spin or photon polarization for a pair of correlated electrons or photons, rather than the properties actually dealt with in the original paper—a practice we will also follow. In the case of such correlated systems, knowledge of, for example, a directional property of one of the two systems implies knowledge, with probability one, of the parallel directional property of the other system, as well as knowledge, with probability less than one, of directional properties of the second system that are not parallel to the directional property of the first system. Since it cannot be the case that *every* measurement on the first system disturbs the second system—in particular, this must fail if the measurement occurs when the two systems are sufficiently far apart in space—two conclusions follow: (1) An explanation of the probabilistic character of the quantum-mechanical description in terms of interference by measurement cannot be right; (2) Since the perfect nature of the correlation holds for *any* of the possible parallel directional properties which we might freely choose to measure, one can hardly avoid the conclusion that the paired system has the directional property which the correlations require it to have with probability one.

By a *disturbance interpretation* of quantum mechanics, let us mean an interpretation which maintains first, that there are properties which measurements uncontrollably disturb, and second, that the uncontrollability of such disturbances explains why the probabilistic description of quantum mechanics cannot be improved upon. It is clear from our brief sketch that the EPR argument contains an objection to a disturbance interpretation's account of the probabilistic character of the quantum-mechanical description of reality and the justification it offers for the claim that the quantum-mechanical description cannot be improved upon. This objection was explicitly drawn by Einstein in a letter to Karl Popper,[16] in which Einstein cites Heisenberg as a (not entirely consistent) representative of what Einstein characterizes as a statistical interpretation that concedes the incompleteness of the quantum-mechanical description but appeals to a disturbance interpretation to argue that the quantum-mechanical description is nevertheless satisfactory, and that the theory is capable of serving as a basis for theoretical physics. Although Einstein finds Heisenberg's interpretation problematic, it is not the primary target of his and EPR's discussions of completeness.

Nor is the primary target of the paper the closely related "statistical" interpretation of Born, who also concedes the incompleteness of quantum mechanics but dismisses the demand for a more complete theory on the ground that such a demand is the result of a metaphysical prejudice. Born's view emerges especially clearly in his correspondence with Einstein and his commentaries on their letters. Born writes of Einstein's letter of September 15, 1950, that Einstein

> calls my way of describing the physical world [by which Born means his celebrated statistical or probabilistic interpretation of the wave function] 'incomplete'; in his eyes this is a flaw which he hopes to see removed, while I am prepared to put up with it. I have in fact always regarded it as a step forward, because an exact description of the state of a physical system presupposes that one can make statements of infinite precision about it, and this seems absurd to me.[17]

And commenting on a letter of December 3, 1953, Born again conflates Einstein's concerns about the *completeness* of the quantum-mechanical description with a preoccupation with its *exactness* and a commitment to infinitely precise measurements; Born concludes by rejecting Einstein's view as "metaphysical nonsense." In his letter to Popper, Einstein dismisses the relevance of what is essentially Born's view with the remark:

> Altogether I really do not at all like the now fashionable [*modische*] "positivistic" tendency of clinging to what is observable. *I regard it as trivial that one cannot, in the range of atomic magnitudes, make predictions with any desired degree of precision.*[18]

The feature of the statistical interpretations of both Heisenberg and Born that Einstein is concerned to emphasize and contrast with the view that *is* his principal target is that, like a disturbance interpretation, Born's and Heisenberg's statistical interpretations *assume* that there are properties which the quantum state specifies only with probability. But Einstein and EPR set out to *show* that there exist properties which are left out of the quantum-mechanical description (and that, according to the disturbance interpretation, measurements disturb), without simply assuming this from the outset. The view that Einstein and EPR are mainly concerned to oppose denies this assumption. That there are such properties is held to follow from EPR's well-known criterion of reality[19] and the fact that they can be predicted on the basis of the quantum theory of correlated systems with probability one. Once this is recognized, EPR argue, the *theory itself* can be appealed to in support of the conclusion that more is true about a system than is contained in its quantum-mechanical description— even if the theory does not predict with probability one the simultaneous values of classically conjugate parameters. EPR infer the incompleteness of the quantum-mechanical description of reality from the theory's failure to include in its characterization of the state of the system all the properties which they take their argument to have shown the system should be understood as having.

Bohr's views are less straightforward. The emphasis on complementarity does not obviously exclude being combined with a disturbance interpretation of the theory. There are, however, numerous places where Bohr explicitly cautions against an understanding of the uniqueness of quantum mechanics along these lines. For example, writing in 1958 in what is perhaps his clearest discussion of the subject, we find the following disclaimer:

> In the treatment of atomic problems, actual calculations are most conveniently carried out with the help of a Schrödinger state function, from which the statistical laws governing observations obtainable under specified condi-

tions can be deduced by definite mathematical operations. It must be recognized, however, that we are here dealing with a purely symbolic procedure, the unambiguous physical interpretation of which in the last resort requires a reference to a complete experimental arrangement. *Disregard of this point has sometimes led to confusion, and in particular the use of phrases like 'disturbance of phenomena by observation' . . . is hardly compatible with common language and practical definition.*[20]

Such remarks suggest that Bohr might be among those who understand the reality of a property of a quantum system as in some sense contingent on its measurement. But the ellipsis conceals Bohr's inclusion of the phrase "creation of physical attributes of objects by measurements" as among those phrases that are "hardly compatible with common language and practical definition." Our quotation is from a relatively late paper. By his inclusion of an explicit rejection of the "creation of physical attributes of objects by measurements," Bohr may well have been influenced by further reflection on the EPR argument and Einstein's views, since only the caution against the assumption of a "mechanical disturbance" appears in the closely similar passage of his original response to EPR.[21] But a fuller understanding of Bohr's position will have to await our reconstruction of the framework of Einstein's and EPR's discussion, and of the difficulties to which it has been discovered to be subject.

In the penultimate paragraph of their paper, EPR single out an assumption of their argument (1935, p. 780; italics in the original):

one would not arrive at our conclusion if one insisted that two or more physical quantities can be regarded as simultaneous elements of reality *only when they can be simultaneously measured or predicted.* On this point of view, since either one or the other, but not both simultaneously, of the quantities P and Q can be predicted, they are not simultaneously real. This makes the reality of P and Q depend upon the process of measurement carried out on the first system, which does not disturb the second system in any way. No reasonable definition of reality could be expected to permit this.

EPR are unapologetic in their dismissal of any response to their argument that would exploit this gap. But many years later, Einstein returned to the issue the assumption raises, and, in his (1949) *Reply,* gave a more careful statement of what is at stake. This discussion was evidently influenced by Bohr's response to the EPR paper, since Bohr is effectively identified as the advocate of the "definition of reality" that EPR had rejected as unreasonable:

> Of the "orthodox" quantum theoreticians whose position I know, Niels Bohr's seems to me to come nearest to doing justice to the problem. Translated into my own way of putting it, he argues as follows:
>
> If the partial systems A and B form a total system which is described by its ψ-function $\psi / (AB)$, there is no reason why any mutually independent existence (state of reality) should be ascribed to the partial systems A and B viewed separately, *not even if the partial systems are spatially separated from each other at the particular time under consideration.* (1949, pp. 681–682, Einstein's italics)

Setting to one side the question of the accuracy of Einstein's characterization of Bohr's position, how are we to understand Einstein's thesis of the mutually independent existence of the two partial systems of an EPR-correlated pair? And what is its relation to the prohibition against action at a distance? The thesis is given a more elaborate statement in his 1948 *Dialectica* article,[22] which appeared a year before the *Reply.* As I understand it, the thesis is informed by two ideas, the first a general formulation of Einstein's realism, and the second, an account of how a philosophical idea of such generality should be understood so that it can be concretely applied in a discussion of the foundations of physics.

Einstein's *realism,* as he explains it in the *Dialectica* article, is based on the notion that things possess what he calls a "being thus," or "state of reality," by which I take him to mean that the fact of their reality is independent of our ability to know them. This is perhaps the essential mark of realism as it has come to be understood in the philosophical tradition.[23] In this respect, Einstein's understanding of realism is closer to the traditional doctrine than it is to scientific realism, which is narrowly focused on whether the theories of physics are true or approximately true. As we saw in our discussion of molecular reality, a preoccupation with scientific realism can easily lead one to understate the basis for existence claims involving the constitution of matter. In the present case, scientific realism lacks an evident connection with the issues that motivate Einstein's interest in the completeness of the quantum-mechanical description of reality. This contrasts with the traditional understanding of realism as the independence of reality from our capacity to know it. By the condition of mutual independence, the traditional notion has an almost immediate application to Einstein's concerns in a way that scientific realism does not. For, in order that the stress which Einstein's realism places on the separation of reality from our possible knowledge of it is have a useful application in the discussion of the interpretation of quantum mechanics, it is necessary to specify the physically important type of knowledge that the reality of things should be understood to be independent of. This point is addressed by the remarks on Bohr in the *Reply* and their further development in the *Dialectica* article. They transform realism from an abstract philosophical claim into a thesis with consequences for *measurement*, and indeed into a thesis with consequences for the measurement of one or another parameter of two correlated but no longer interacting systems. When elaborated and applied to the case of EPR-correlated systems, Einstein's realism is directly opposed to the view he attributed to Bohr: Einstein's realism requires that the values of a parameter (or the holding of a property) of one of two correlated systems cannot be affected by the *kind* of measurement performed on the system with which it is paired—nor can it be affected by *whether* a measurement is or is not performed on the paired system—once the interaction has ceased and the two systems are sufficiently far apart in space.

What we will refer to as Einstein's *local realism* is just the formulation of his realism given above together with its elaboration in terms of this understanding of the mutually independent existence or being thus of two previously interacting, but now no longer interacting, spatially separated systems. Notice that local realism is a condition that is imposed at the operational or "surface" level of measurable parameters, and it is in this respect like the no-signaling condition formulated in Bub (2016, p. 76). We will return to this point in the next section.

Einstein's goal, and the goal of the EPR paper, is to show the incompatibility of local realism with the view that quantum mechanics is complete. The difference between Einstein's and EPR's presentations is partly one of emphasis: for EPR the criterion of reality is isolated and accorded a central position in their argument, while in the case of Einstein, local realism and its analysis dominate his discussion. Despite such differences, it is clear from his letter to Popper that Einstein accepted the argument of the EPR paper and that he concurred with its goal of establishing the incompatibility of the criterion of reality with the completeness of quantum mechanics. However, the letter also shows him to have taken EPR's conclusion a step further by combining it with his local realism to argue for an epistemic (or, as Einstein puts it, "statistical") interpretation of the wave function, and perhaps— although this is an extrapolation—for the program of completing the theory along the lines of what I have characterized as the Einsteinian response. In this letter and in the *Reply*, Einstein takes the development of such an epistemic understanding of the wave function to be the evident lesson of the possibility of assigning many different wave functions to a system on the basis of measurements carried out on the system with which it is paired.[24] This view of the significance of the EPR argument for the understanding of the wave function is contrasted with the position Einstein attributes to Bohr: By denying local realism, Bohr is free to argue that the quantum-mechanical description is complete *as it stands,* since there is nothing "in reality" that it can be said to have missed.

In his *Dialectica* paper, Einstein also formulated a *principle of local action:*

> For the relative independence of spatially distant things (A and B), this idea is characteristic; an external influence on A has no *immediate* effect on B.[25]

Einstein remarks that this principle is applied consistently only in field theory. He then adds that "[f]ield theory has carried [it] out . . . to the extreme, in that it localizes within infinitely small (four-dimensional) space-elements the elementary things existing independently of one another that it takes as basic, as well as the elementary laws it postulates for them."[26] But although Einstein was evidently partial to field theory, there is no suggestion that the introduction of the principle of local action is a disguised argument for a field-theoretic approach, still less for an approach that takes as basic, "infinitely small (four-dimensional) space-elements."

The point of Einstein's introduction of the principle of local action receives some further clarification in the very next paragraph when he remarks of the principle of local action that its "complete suspension would make impossible the idea of the existence of (quasi-) closed systems and, thereby, the establishment of empirically testable laws in the sense familiar to us."[27]

This statement echoes an earlier comment (made prior to the introduction of the principle of local action) that was advanced in support of his local realism: "Without such an assumption of the mutually independent existence (the 'being-thus') of spatially distant things, an assumption which originates in everyday thought, physical thought in the sense familiar to us would not be possible."[28] Taken together, these remarks might appear to justify the equation of local realism with the principle of local action and the prohibition of action at a distance. But to equate local realism with the principle of local action would give a misleading representation of the relation between these two theses.

Local realism receives support from the principle of local action and the rejection of action at a distance, but it has a more fundamental and even conceptual status than either of these principles. When Einstein introduces his local realism, he does so by locating it among ideas that are characteristic of "the realm of physical ideas independently of the quantum theory" that have their origin in everyday thinking (ibid). Having this fundamental status, it would be surprising if local realism were not satisfied by *Newtonian* mechanics, even if the theory of Newtonian gravitation assumes the existence of an action at a distance force. Not only is Newtonian mechanics one among the ideas of physics independently of the quantum theory, but the notion that there are quasi-closed systems of the sort that local realism requires is compatible with an action at a distance theory like Newton's *just because that theory allows for the kind of approximative reasoning that the concept of a quasi-closed system rests upon.* Local realism is therefore a constraint on theory construction that is satisfied by Newtonian mechanics and by the conception of theory testing that emerged from the Newtonian tradition of theory-mediated measurement.* For Einstein the physical idea with which it is most closely associated is *not* the denial of action at a distance, but the rejection of action at a distance theories that preclude quasi-closed or isolated systems. Although Einstein's allusion to "spooky actions at a distance"[29] is invariably represented as an outright rejection of action at a distance theories like Newton's, it is clearly compatible with the more nuanced rejection of only those action at a distance theories which would preclude the notion of a quasi-closed system. More importantly, and independently of

---

*Editor's note: The ambiguity of "modern" in the editor's note to the first page of the Introduction of this volume concerns whether both Newton's theory (of gravity) and *relativistic* gravity (general relativity) satisfy the same constraints, or whether only what eventually emerged from the Newtonian tradition satisfies the relevant constraints. It turns out that the Newtonian *methodology* is compatible with both; see the Editor's Foreword.

the question of Einstein interpretation, Einstein was evidently correct to separate the two theses, since for theories which deny the possibility of quasi-closed systems, the principle of local action would be "completely suspended," and in this case, there would also be a conflict with local realism.

The "orthodox quantum theoreticians"—by which Einstein meant those physicists who followed Bohr in their justification of why quantum mechanics is complete—are committed to an understanding of the theory as one in which the principle of local action is "completely suspended." Local realism figures centrally in Einstein's objection to what he understood to be Bohr's proposal for understanding quantum mechanics so that, despite its statistical character, it is nevertheless complete. According to this proposal the EPR argument involving separated systems does not call into question the completeness of the quantum-mechanical description of reality because *there is no real factual situation at the distant location until an intervention has been made involving the system with which it is paired*. Einstein rejected quantum mechanics under its "orthodox" interpretation because such an interpretation makes the theory utterly anomalous among extant physical theories.

4.3   Bell's Theorem and Einstein's Local Realism

Einstein maintained, on the basis of the EPR paper and closely related considerations, that the quantum-mechanical description of the EPR correlations is incomplete because of its failure to include a description of the "real factual situation" of each of the two correlated systems. In this section, we will consider the possibility of completing the quantum-mechanical description in accordance with Einstein's local realism.

A great deal of attention has been devoted to showing where EPR's argument against the completeness of the quantum theory fails, but by far the most important objection to the argument is the one raised by Bell's theorem.[30] Bell focused on the possibility of what he characterized as a complete, *locally causal* account of the EPR correlations. His analysis revealed that such an account of the correlations leads to a condition of "factorizability." He then proved a theorem showing that factorizability implies an inequality that is incompatible with the quantum-mechanical probabilities exhibited by these correlations. Bell's discussion is of special interest to us because of its bearing on Einstein's local realism and the difficulty of combining it with an account of the correlations that is more "complete" than the one given by quantum mechanics.

Before turning to Bell's analysis of local causality, let me conclude this preliminary discussion with a formulation of Bell's theorem that is due to Itamar Pitowsky.[31] Pitowsky's formulation will prove to be of fundamental importance when we come to consider the significance of the EPR correlations for understanding the basic conceptual innovation of the quantum theory.

Let F be a finite set $\{A, B, C, \ldots\}$ of Hermitian operators acting on a finite-dimensional Hilbert space $\mathbf{H}$. The operators in F represent a family of dynamical variables with possible values corresponding to the eigenvalues of the operators. A quantum state ρ assigns a probability distribution P_ρ to the possible values of the dynamical variables represented by the operators in F. For each possible value $a_1, a_2, \ldots$ of A we have $P_\rho(a_1|A)$, $P_\rho(a_2|A)$, $\ldots$, i.e., $P_\rho(a|A)$ is the probability ρ assigns to the property, A's having the value a, or equivalently, to the proposition that A has the value a. Similarly, for each value b_1, b_2, ... of B we have $P_\rho(b_1|B)$, $P_\rho(b_2|B)$, ..., etc. (The set F may be so chosen that the probabilities associated with this finite set completely characterize ρ as a state on H, but we make no use of this fact here.)

As observed by Kochen and Specker, there always exists a classical probability measure (the product measure),

$$P_\rho(a, b, c, \ldots \mid A, B, C, \ldots) = P_\rho(a|A)P_\rho(b|B)P_\rho(c|C) \ldots,$$

on the operators in F, provided we ignore algebraic relations among them and assume that the dynamical variables they represent are all independent of one another.

Suppose, however, that we require of P_ρ that it satisfy the following two conditions:

(1) $P_\rho(a, b, c, \ldots \mid A, B, C, \ldots)$ is a probability measure defined on all eigenvalues of all the operators $A, B, C, \ldots$ in F;
(2) If $A, A', A'', \ldots$ in F *commute*, then $P_\rho(a, a', a'', \ldots \mid A, A', A'', \ldots)$ coincides with the quantum-mechanical probability assigned by ρ.[32]

Then *Bell's theorem* tells us that there is an F and a ρ such that P_ρ does not exist; in this case, the operators in F are local operators acting on a tensor product of Hilbert spaces; and *Kochen and Specker's theorem* tells us that there is an F such that for *any* ρ, the distribution P_ρ does not exist.

Pitowsky[33] has shown that the existence of P_ρ is equivalent to the requirement that the numbers

$$\{P_\rho(a, a', a'', \ldots \mid A, A', A'', \ldots): A, A', A'', \ldots \text{ in } F \text{ commute}\}$$

satisfy a finite family of linear inequalities, which George Boole called "conditions of possible experience." The nonexistence of P_ρ implies the failure of at least one such inequality. Bell's original formulation of his theorem consisted in the derivation of such an inequality for the case where ρ is the EPR state. The requirement that P_ρ should satisfy a certain finite system of linear inequalities like the one derived by Bell is equivalent to the requirement that P_ρ should be representable as a weighted sum of two-valued measures.[34]

Hence, although the theorem of Bell and that of Kochen and Specker make very different claims regarding F, ρ, and P_ρ, they both expose the incompatibility of the probabilities of quantum mechanics with the special status that two-valued measures enjoy in classical probability theory—probability theory understood as a part of the theory of measure over a Boolean algebra.

To briefly recapitulate, even if there exists what Einstein and EPR would regard as a complete account of the marginal probabilities and the perfect correlations of the EPR state for *parallel* directional properties, this is less than what EPR need to establish the conclusion that quantum mechanics is incomplete. It is also necessary to show that it is possible to have an account within classical probability theory that encompasses the conditional probabilities of *all* pairs of propositions involving the directional properties of the two correlated systems—including those propositions which contain directional properties that are *not* parallel. Bell's theorem implies that EPR's understanding of the 0–1 conditional probabilities of parallel directional properties and their measurement cannot be extended to the "off" 0–1 conditional probabilities associated with directional properties of the two systems that are not parallel; hence, the quantum-mechanical probabilities associated with all pairs of directional properties are not representable as weighted averages over a collection of two-valued measures. By the correspondence between two-valued measures and two-valued homomorphisms or truth-value assignments, such an account of the correlations is naturally interpreted as a necessary component of a completion of the quantum-mechanical description by one that records what Einstein understands as *the real factual situation* of each of the paired systems: The real factual situation corresponds to all those propositions that are mapped onto "truth" by a truth-value assignment. It is transparent from Pitowsky's formulation that Bell's theorem shows that, even if we set to one side Einstein's concern with recording the real factual situation at each location, the idea that the quantum theory of a pair of correlated systems can be understood against the background of classical probability theory is called into question because it is based on too restrictive a metatheoretical framework to encompass theories which employ probability assignments of the kind we encounter in cases like the EPR correlations.

Thus far we have focused on Bell's *theorem* and its relation to classical probability theory and a plausible construal of Einstein's notion of the real factual situation of the two component particles. Let us now to turn to Bell's *analysis* of local causality and its bearing on Einstein's local realism.[35]

Although the framework of Bell's discussion does not necessarily preclude the possibility of a preferred foliation of space-time—i.e., a foliation associated with a privileged inertial frame—it does assume a significant amount of causal structure that we associate with Minkowski space-time: The spatiotemporal framework preserves the separation of regions effected by the light cone structure of Minkowski space-time; the causal connectibility of events is understood in terms of the possibility of sending a massive or massless particle from the location of one event to that of the other event; the causal relation among events agrees with the relation of temporal precedence, so that e is a possible cause of f only if e temporally precedes f; and the relation *neither causally precedes* is not transitive. The causal structure provides the conceptual background for Bell's assessment of the prospects for a locally causal and "appropriately complete" account of the EPR correlations.

Bell's discussion of the EPR correlations is expressed in the framework of an experimental analysis, rather than in terms of directional properties or dynamical variables. At the surface level of experimental results, we imagine two spatially separated experimental arrangements, each consisting of a polarizer and a counter. We denote by A and B the outputs of the counters at the two spatially separated locations. A and B take values +1 (Yes) or –1 (No) according to whether or not a photon passes through the polarizer to which the counter is connected. The polarizers are set at various angles a and b, relative to some standard direction in which they are parallel. On this analysis, the measurement outcomes recorded by the counters replace the values of dynamical variables of the algebraic analysis; hence, instead of the notion of a directional *property* of photons, we consider the directional *settings* of the polarizers. This "operationalizes" the discussion, but its point is not to give a reductive analysis of the correlations. The discussion aims, rather, to minimize conceptual commitments in order to make the analysis free of assumptions that might be seen as prejudicial to a locally causal account; this applies especially to assumptions that derive from the mathematical structure of Hilbert space and that dominate algebraic analyses of the problem of hidden variables. Bell's goal is to represent the correlations as empirically given *phenomena* and to investigate the possibility of a complete, locally causal account of them.

Moving from the surface level of counter outputs and polarizer settings to the level of *underlying parameters*, Bell introduces the variables c and λ. The variable c ranges over any relevant parameters in the causal pasts of the settings of the polarizers and the recording of outputs by the counters. More precisely c ranges over parameters of a spatiotemporal region that crosses the backward light cones of the polarizers and counters in an area in which their backward light cones do not overlap. And λ ranges over whatever parameters are necessary to complete the quantum-mechanical description "in the way envisaged by EPR." That notion of completeness may correspond to the existence of a truth-value assignment along the lines of our discussion of Bell's theorem, but Bell's discussion does not assume this: Bell's use of "completeness" is pre-analytic and intuitive. This is in keeping with the theoretical neutrality of Bell's description of the correlations as empirically given phenomena.

Given these preliminary expository remarks, Bell explains the overall logic of his argument as follows:

Let

$$\{A, B \mid a, b, c, \lambda\}$$

denote the probability of particular values A and B, given values of the variables listed on the right. By a standard rule, the joint probability can be expressed in terms of conditional probabilities:

$$\{A, B \mid a, b, c, \lambda\} = \{A \mid B, a, b, c, \lambda\} \{B \mid A, a, b, c, \lambda\}.$$

Invoking local causality, and the assumed completeness of c and λ in the relevant parts of [the spatiotemporal region that crosses the backward light cones of the two spatially separated regions], we declare redundant certain of the conditional variables in the last expression, because they are at space-like separation from the result in question. Then we have

$$\{A, B \mid a, b, c, \lambda\} = \{A \mid a, c, \lambda\}\{B \mid b, c, \lambda\}. \qquad (10)$$

Now this formulation has a very simple interpretation. It exhibits A and B as having no dependence on one another, nor on the settings of the remote polarizers (b and a respectively), but only on the local polarizers (a and b respectively) and on the past causes, c and λ. We can clearly refer to correlations which permit such factorization as "locally explicable."[36]

Thus, on Bell's explication, Einstein's local realism requires a locally causal and complete description of the correlations; and, by the above argument, (10) is a necessary component of such a description in terms of the underlying parameters c and λ. Recall from our discussion in Section 4.2 that Einstein's realism requires a condition of "mutual independence": The value of a parameter (or the holding of a property) of one of two correlated systems cannot be affected by the *kind* of measurement performed on the system with which it is paired—nor can it be affected by *whether* a measurement is or is not performed on the paired system—once the interaction has ceased and the two systems are sufficiently far apart in space. As we remarked, Einstein's local realism is a constraint that is imposed at the surface or operational level of the measurable parameters that characterize the two component systems and that are related by the correlations situating Einstein's mutual independence condition in the context of a locally causal and complete theory. On the assumption that a locally causal completion should appeal only to parameter values prior to a measurement of a directional property, and assuming that the setting of each of the polarizers can be freely chosen by the experimenter,[37] knowledge of the underlying parameter values should afford an explanation of the correlations that traces their source to the values that the underlying parameters assume in the disjoint union of appropriate regions of the causal pasts of the paired systems. *But then knowledge of the underlying parameter values must "screen off" or make redundant any information that the measurement of a directional property of the first system can contribute to the conditional expectation of a measurement of the corresponding directional property of the second system.* As Bell shows, any explanation in terms of underlying parameters that satisfies these conditions must also satisfy a condition of factorizability. By Bell's theorem, this in turn implies that the correlation involving parallel properties cannot be extended to a locally causal account that encompasses all of the probabilities exhibited by the correlations of the EPR state. On Bell's analysis, factorizability is not an assumption of the argument against local realism, but a consequence of the analysis of what is required when local realism is extended to the underlying parameters of a locally causal *completion* of the quantum-mechanical account of the EPR correlations. Hence its justification does not appeal to assumptions about the "separability of properties" as Howard (1985, 1989) has maintained, or to assumptions that are peculiar to Einstein's "field-theoretical program," as Teller (1989) has argued. Such considerations are as extraneous to Bell's analysis as they are to Einstein's understanding of local realism. As Bell noted, "[v]ery often such factorizability is taken as the starting point of the analysis. Here we have preferred to see it not as the *formulation* of 'local causality,' but as a consequence thereof."[38]

We have seen that in his analysis of local causality, Bell relies on only a pre-analytic and nontechnical sense of "completeness" when he requires that events in the backward light cones of each of the paired systems must be specified completely. This sense of "completeness" suffices to exclude the possibility that the traces of events in these regions might supplement whatever else might be used for calculating the probability of an outcome of a measurement on one system, conditional on a measurement of the system with which it is paired: Completeness ensures that any local causal influence has been "screened off," or "shielded against," in Bell's terminology. Shielding implies that factorization is justified as a premise in the derivation of the Bell inequality. But although the notion of completeness that enters into Bell's *analysis* of local causality is a nontechnical one, a notion of "classical" completeness is an essential part of the content of Bell's theorem. This emerges from Pitowsky's formulation of the theorem, which is based on the observation that the satisfaction of a Bell inequality by a probability measure P_ρ is *equivalent to P_ρ* being representable as a weighted sum of two-valued measures. The equivalence follows from Pitowsky's account of the EPR, and other correlations, by the notion of a correlation polytope, and the duality of the representation of such polytopes in terms of their facet inequalities (such as the Bell inequality) or in terms of their extremal points (two-valued measures, in the case of the correlation polytope corresponding to the EPR correlations).[39] We can therefore conclude from the theorem establishing Bell's inequality that EPR require that the probabilities exhibited by the

EPR state should be representable as weighted averages of two-valued measures. But this is the paradigm of a classical completeness condition. The combination of Bell's analysis of local causality together with Bell's theorem therefore yields an argument against the idea that the quantum-mechanical description of reality can be completed by a locally causal and *classical* theory, where, by a *classical theory* we mean one whose probability measures are representable as averages of two-valued measures. By exploiting the correspondence between two-valued measures and two-valued homomorphisms, this conclusion can be taken a step further, and it can be argued that EPR require that the probabilities of the EPR state must be representable as weighted averages of *truth-value assignments.* On the assumption that a truth-value assignment captures Einstein's idea of a "real factual situation," it follows that Bell's analysis of local causality together with Bell's theorem argues against the possibility of completing the quantum-mechanical description of reality by a locally causal theory in which probabilities are represented as averages over real factual situations.[40]

The justification of Bell's use of factorizability—or *conditional statistical independence*—can be clarified by the distinction between parameter and outcome independence. *Parameter independence* asserts the independence, relative to λ and c, of the probability of the outcome of a measurement, at one location, of the *orientation* of the polarizer at the other location; and *outcome independence* asserts the independence, relative to λ and c, of the probability of the outcome of a measurement at one location, of the *outcome* of a measurement at the other location. As is well known, conditional statistical independence is equivalent to the conjunction of parameter and outcome independence. Parameter independence is conceptually very closely related to "the no-signaling condition"; indeed, the no-signaling condition is just parameter independence without the relativity to the underlying parameters λ and c.[41] Like Einstein's mutual independence condition, no-signaling applies at the surface or operational level of measurement results. It differs from Einstein's requirement of mutual independence by formulating a requirement on marginal *probabilities* rather than parameter *values*: According to *the no-signaling condition,* the marginal probability for the directional properties of either of two correlated systems cannot be affected by the *kind* of measurement performed on the system with which it is paired—nor can it be affected by *whether* a measurement is or is not performed on the paired system—once the interaction has ceased. Parameter independence simply *extends* the no-signaling condition—which, like Einstein's mutual independence condition, concerns the surface or operational level of the measurement of directional properties—to a condition which holds at the level of the underlying parameters that might be introduced to *complete* the quantum-mechanical description of the correlations we observe at the operational level of the measurement of directional properties.

Examples of stochastic hidden-variable theories that exhibit parameter independence and outcome *dependence* have attracted some interest because without outcome independence the derivation of a Bell inequality is blocked. Given its close connection to the no-signaling condition, it is natural to suppose that a relevant notion of locality is captured by just parameter independence. It can then be argued that a stochastic hidden-variable theory which satisfies parameter independence is an appropriately local theory even though it does not yield a locally *causal* account of the correlations. The theories envisaged by this approach treat the correlations as irreducibly stochastic, and so, not causally explicable—whether in terms of local or nonlocal causes.[42] Such *theories* may be local, because they satisfy parameter independence, even though the *correlations* they allow are nonlocal in the sense that they hold independently of the distance separating the correlated systems.

Whatever other interest it may have, the possibility of a stochastic theory which exhibits parameter independence but outcome dependence is largely irrelevant to Einstein's conception of local realism and the bearing of Bell's analysis on the EPR argument against the completeness of quantum mechanics. Einstein did not object to the quantum theory merely because it is a statistical theory. He objected to the idea that as a statistical theory quantum mechanics might nevertheless serve as the basis of theoretical physics.[43] But preserving parameter independence while rejecting outcome independence in order to allow for an irreducibly stochastic theory of the correlations only succeeds in replacing one statistical theory by another. By its very nature, such a theory leaves the correlations unexplained, and for this reason, it cannot, in Einstein's view, serve as the basic theory of physics any more than the "statistical" quantum theory. This objection to stochastic hidden-variable theories does not invoke Einstein's local realism. And indeed, as we have seen, Einstein's appeal to local realism is not part of an objection to the quantum theory, but to a certain interpretive proposal for maintaining the theory's completeness.

As a contribution to the evaluation of hidden-variable alternatives to quantum mechanics, Bell's analysis of the problem of completing the quantum-mechanical description along the lines envisaged by Einstein's local realism is very compelling; it is independent of the algebraic structure of the quantum theory, and it requires only the recovery of the empirical predictions to which the theory leads us. Given certain generally accepted assumptions about the causal structure of space-time, on Bell's analysis, the experimental predictions *by themselves* resist an explanation that conforms to Einstein's local realism. By contrast, Kochen and Specker's approach to the problem of hidden variables explicitly relies on the algebraic structure that the theory ascribes to the possible properties of a physical system. Kochen and Specker assume—without an extended argument or analysis—that a hidden-variable theory should be required to preserve this structure.

Even if we reject Kochen and Specker's approach to hidden-variable alternatives to quantum mechanics, their results remain important for what they reveal about the status of the theory's algebraic structure. If the algebra of properties that quantum mechanics associates with a physical system were embedable into a Boolean algebra, it might reasonably be argued that the contribution of the algebraic structure is merely pragmatic and convenient, rather than factual and principled. The non-embedability of the quantum algebra of observables into a Boolean algebra— and, a fortiori, of the quantum algebra of properties into a Boolean algebra—blocks this interpretation of its significance for the theory. The algebraic structure of the theory—and the probability assignments to which it leads—is arguably the theory's distinctive conceptual innovation.

Kochen and Specker's theorem, showing that there exists a finite family of possible properties of a quantum-mechanical system that has no classical truth-value assignment, raises the question of the appropriate notion of completeness in a foundational discussion of the quantum theory. The theorem suggests that the claim that the quantum theory is complete should be understood as a claim about the "completeness" of the set of its "statistical" states. On this analysis, for the theory to be *complete,* the set of quantum-mechanically generated statistical states must contain every positive real-valued measure on the algebra of properties of a quantum-mechanical system that is a classical probability measure on the Boolean subalgebras of the algebra. Gleason's (1957) theorem tells us that the theory is complete in this sense, even though its statistical states are not, in general, representable as weighted averages of two-valued measures—as they are in classical theories. In the evidentiary framework, probability is conceptualized entirely classically: When we consider many repetitions of an experiment in order to evaluate the frequency of occurrence of a particular measurement outcome, we take a ratio of outcomes that is favorable to a theoretical prediction of its probability to be confirmatory of the theory.[44] But although the classical conception of probability is applicable to the evaluation of measurements involving a single property (or collection of properties associated with a family of mutually commuting observables—a qualification I will henceforth implicitly assume), the theoretical framework that comprises the quantum theory's probabilistic predictions for all possible properties has a radically different character, one that does not admit a description in terms of the classical conception of probability.

In light of the completeness of the theory, established by Gleason's theorem, the principle by which the quantum theory generates its probabilistic predictions is clearly of theoretical significance. The geometrical character of this principle is especially evident in the case of the EPR correlations. Here the relative angles of directions in physical space by which the spin or polarization properties are distinguished are encoded in Hilbert space and in the principle by which a probability is assigned to one directional property, conditional on another's having been assigned probability one. Bell's theorem shows that in the case of the EPR state, the totality of these theoretically derived probability assignments cannot be interpreted as ratios of classical truth-value assignments. Rather they reveal a conception of probability that separates the conceptual connection between probability and truth. This contrasts with the evidentiary framework, where it is assumed that probabilities can always be conceptualized entirely classically, as the ratios of possible truth-value assignments.

The present "algebraic" or "probabilistic" approach to the central conceptual innovation of the theory suggests an account of Bohr's reply to EPR that is very different from the position attributed to him by Einstein. Recall that Einstein understood Bohr to have denied the mutual independence of the subsystems involved in the EPR correlations and to have held that there is no real factual situation associated with either subsystem until a measurement interaction has taken place involving one of the paired subsystems. As EPR remarked (Einstein, Podolsky, and Rosen 1935, p. 780), such a position "makes the reality of [the subsystem's properties] depend upon the process of measurement carried out on the first system, which does not disturb the second system in any way. No reasonable definition of reality could be expected to permit this." But this unnecessarily saddles Bohr with a commitment to a philosophical thesis regarding the real factual situation in Einstein's sense. Give the persistent ambiguity in Bohr's formulations of complementarity between indeterminism and indeterminacy—an ambiguity we noted at the beginning of Section 4.2—Bohr should be understood to have set aside Einstein's question about the real factual situation and to have pursued the question of whether it is possible to extend EPR's account of the marginal probabilities and the perfect correlations to one that encompasses *all* the quantum-mechanical probabilities that are exhibited by the EPR state. In doing so, Bohr was indirectly appealing to a conception of completeness that transcends Einstein's particular focus on the real factual situation, one that was only fully clarified with the formulation and proof of Gleason's theorem.

What Bohr explicitly says in his original response to EPR is very suggestive. Bohr observes that without an extension of their argument to "the . . . conditions which define the possible types of predictions regarding the future behavior of the system," the conclusion that the quantum-mechanical description is essentially incomplete is not justified. Here is the full passage from which this remark is taken:

> Of course there is in a case like that just considered no question of a mechanical disturbance of the system under investigation during the last critical stage of the measuring procedure. But even at this stage there is essentially the question of *an influence on the very conditions which define the possible types of predictions regarding the future behavior of the system.* Since these conditions constitute an inherent element of the description of any phenomenon to which the term "physical reality" can be properly attached, we see that the argumentation of the mentioned authors does not justify their conclusion that quantum-mechanical description is essentially incomplete.[45]

The understanding of this passage turns on the explication of the nature of the "influence on the very conditions which define the possible types of predictions regarding the future behavior of the system," and of the claim that these conditions "constitute an inherent element of the description of any phenomenon to which the term 'physical reality' can be properly attached." I take the focus of this passage to be the quantum-theoretical derivation of probabilistic predictions about the future behavior of a physical system. Bohr's point is that these predictions are derived from "conditions" that are inherent in a theoretical description of the phenomena that captures something physically real. But what are these conditions? From our perspective, the conditions isolated by Bohr are given by the theory's representation of the algebraic structure of the possible properties of a system, since this plays a fundamental role, and indeed *defines*—in the sense that it *uniquely determines*—the theory's probability assignments. This structure constitutes the key principle that distinguishes quantum mechanics from classical mechanics. On this explication, the fault in EPR's discussion lies in their not having appreciated the full extent of the theoretical involvement of the algebraic structure in the derivation of the theory's probability assignments. In the case of Einstein, this is especially ironic, given the centrality of the principle-theoretical components of theories to his widely quoted semi-popular remarks about the nature of the innovations of the theory of relativity.[46]

Subsequent developments have made it possible to formulate the importance of the algebra of properties of a quantum system for understanding the probabilistic character of the theory more clearly and more definitively than Bohr was able to do. Unlike Bohr, we can appeal to the *completeness* that Gleason's theorem established regarding the set of the theory's statistical states, and to Bell's theorem, showing that the *incompleteness,* which EPR's concentration on just the marginal and 0–1 conditional probabilities suggests, is illusory. Bell's theorem shows that our classical expectations regarding completeness are misplaced, while

Gleason's theorem directs us to the concept of completeness that is appropriate for a probabilistic theory like quantum mechanics.

By comparison with Bohr, Einstein's philosophical views about quantum mechanics require little interpretation or reconstruction. Einstein consistently argued for the development of a theory which would recover quantum mechanics, together with its probabilistic predictions, from a "description of the real factual situation," while clearly preserving the mutual independence of spatially separated and no longer interacting systems. If by a "description of the real factual situation" we understand one that arises within the representation of the possible properties of a system of the sort that is familiar from classical physics—or more generally, from classical probability theory—then this appears to be excluded. By Kochen and Specker's theorem, we cannot recover the *theoretical structure* of quantum mechanics from a commutative theory, and by Bell's theorem, we cannot recover the probabilistic *predictions* of quantum mechanics in the implicit probability theory of classical physics. Only such a theory is capable of meeting Einstein's requirements. In light of these results, quantum mechanics marks a unique departure from our prior experience with conceptually revolutionary transitions from the theories of classical physics.

4.4   Quantum Mechanics and Reality

It was clear almost since its inception, but especially in light of Minkowski's contribution, that special relativity was about a new conception of the structure of space-time. On this understanding of the significance of the special theory, the phenomena of length contraction and time dilation are accounted for by the transition from a theory of space-time in which spatial and temporal distances have an absolute significance, to a spatiotemporal framework in which these notions are no longer absolute, but where contraction and dilation are traceable to the structure of space-time rather than the material constitution of rods and clocks. The space-time perspective differs from Lorentz's recovery of universal Lorentz invariance from an underlying theory of the constitution of matter as basically electromagnetic, a theory whose symmetry group is therefore the symmetry group of Maxwell's theory.[47] What is distinctive about the Einstein-Minkowski view is that it does not recover the universality of Lorentz invariance from an underlying theory of the *constitution of all matter,* but takes the Lorentz group to be the appropriate relativity group regarding *all motions*—whatever the constitution of what is moving—with the consequence that the structure of space-time is represented as Minkowskian rather than Newtonian. The orthodox view of the theory as a theory of space-time structure came about remarkably early in the evolution of our understanding of special relativity.[48]

Nothing comparable has attended our understanding of the nature of the departure of quantum mechanics from classical modes of thought. The proposal we have advanced for understanding the novelty of the quantum theory assimilates the conceptual shift that quantum mechanics exemplifies to the kind of change with which we are familiar from special relativity. The quantum theory introduces the concept of a probability that is based on a change in the structure of the algebra of properties on which probabilities are defined, a change from the structure of a Boolean algebra to an algebra of properties exhibited by its Hilbert space representation. The theoretical understanding of all the peculiarities of the probabilities with which the quantum phenomena present us rests on novel features of this algebraic structure, just as, in special relativity, the phenomena of time dilation and length contraction trace back to novel features of the structure of Minkowski space-time. This is the sense in which, on the present proposal, the nature of the transition from classical mechanics, like that from Newtonian mechanics to relativity, is principle-theoretic rather than constructive. Both transitions involve a reconceptualization of a fundamental principle-theoretic assumption—involving geometric structure in one case, and algebraic structure in the other—rather than a reconceptualization of a constructive-theoretic component of our understanding of the constitution of matter.

How are we to understand probability in quantum mechanics if, as we have argued, the quantum-mechanical concept is fundamentally different from the classical one? Pitowsky (2006) has shown that the epistemic concept, according to which the principles of probability are justified by their coherence, carries over to a justification of all but one of the principles which characterize the quantum-mechanical concept. The exception is a principle that is required by the fact that the algebra of properties—in the language of probability theory, the "space" of "events" to which probabilities are assigned—is not a Boolean algebra, but is the generalization of a Boolean algebra of properties that is effected by Hilbert space.

For our immediate purposes, there are two especially salient features of this generalization: The generalization is representable as a family $\mathbf{F}$ of Boolean algebras with the property that the intersection of any two algebras in the family is also an algebra of the family; and there are events e, represented by minimal non-zero elements of the algebra (algebraic atoms), that belong to algebras B_i and B_j in $\mathbf{F}$, but there is no algebra B_k in $\mathbf{F}$ that contains

Bi and *B_j*. The generalization assumes that *e* is the same property or event whether it is regarded as a member of *B_i* or of *B_j*. We will refer to this feature as the *noncontextuality* of the quantum-mechanical representation of the possible properties of a physical system, or equivalently, as the noncontextuality of its representation of the space of events.

It is characteristic of the epistemic concept of probability that it leads to a purely normative justification of the axioms of the classical theory. On the epistemic interpretation of probability, the justification of the axioms is normative because, among other things, it assimilates the status of the probability axioms to the principles of logic: If our reasoning is not in accordance with the principles of logic, we have no assurance that we will not be led to a false conclusion from true premises. Adherence to logical principles assures us of the *soundness* of our reasoning. According to the epistemic justification of the axioms of probability, our betting on the outcomes of a game of chance will incur a sure loss if our bets do not conform to the classical probability axioms. Arranging our rational expectations in conformity with the axioms of probability assures us of their *coherence.*

Following Pitowsky, in a quantum probability framework, betting behavior is not restricted to a single Boolean algebra of possible outcomes, but involves a finite family of Boolean algebras that do not themselves form the basis of a representation of a single Boolean algebra. In slightly greater detail, Pitowsky asks us to imagine a family of "games," each of whose possible outcomes generates a Boolean algebra, and to consider only families of games whose associated algebras have the character of a family of Boolean algebras that represent the structure of the properties of a quantum-mechanical system. Pitowsky associates with such a family a *quantum gamble.* In a quantum gamble bets are placed on the possible outcomes of each game in the family. In order that our bets should conform to the probabilities of quantum mechanics, it is necessary that they should respect the noncontextuality of the quantum-mechanical representation of events; this requires that our bets should assign the same probability to an event e in each of the games (of the gamble) of which e is a possible outcome, even though the possible outcomes of a gamble may belong to different games—different Boolean algebras of events. A gamble is "taken" when the "croupier" chooses one game from the family and returns all bets on the other games, after which the selected game is played.

In Pitowsky's quantum gambles, unless one's bets conform to a possible quantum-mechanical assignment of probabilities, one can be assured of incurring a loss in a "larger" quantum gamble. This is the content of his Corollary 7 (Pitowsky 2006, p. 227) which Pitowsky labels "completeness": Let F_0 be the set of all the algebraic atoms belonging to the games of the family $\mathbf{F}_0$ of a particular gamble, and let p_0 be a probability assignment to the atoms in F_0 that contradicts every possible quantum-mechanical assignment of probabilities to these atoms. Then there is a finite family $\mathbf{F}$ of games which includes $\mathbf{F}_0$ and is such that p_0 cannot be extended to a quantum-mechanical probability measure on the set F of algebraic atoms belonging to the games in $\mathbf{F}$. When we recall that a quantum-mechanically generated probability assignment (or "statistical state") is a classical probability assignment on Boolean subalgebras of the algebra of events, the corollary tells us that any probability assignment to the algebraic atoms of $\mathbf{F}_0$ that contradicts all possible quantum-mechanical probability assignments must fail to be coherent on some finite extension of the gamble.

Quantum probabilities are therefore like classical probabilities in being susceptible to a "Dutch book" justification, but it is only because of the empirical success of quantum mechanics that the principle of noncontextuality can be regarded as a reasonable constraint on our betting behavior. The justification for betting in conformity with noncontextuality rests on the conviction that, by virtue of its recovery of the observed relative frequencies, quantum mechanics has isolated the true nature of the algebraic structure of the properties associated with a physical system. The theory of probability implicit in quantum mechanics is therefore not a purely normative theory, the source and justification of whose principles consists entirely in the coherence they give to our rational expectations. Nor is it like logic in having an a priori justification. The theory contains the principles of the classical theory, but it is distinguished from it by the presence of the principle of noncontextuality. Unlike the principles of classical probability, the truth of noncontextuality rests on more than a purely normative justification. The source of the principle and its justification derive from the truth of the empirical theory that introduced it. And this gives the concept of probability in quantum mechanics a physical content that would be missed on a purely epistemic interpretation of the concept.

The notion of a quantum gamble illustrates how quantum probabilities, like classical probabilities, are manifested in relative frequencies—albeit in a restricted sense. By hypothesis, the physical system employed in a quantum gamble is used just once, and is then discarded or destroyed. Nevertheless, it is possible to record the frequency of outcomes of a gamble by repeating any *one* game belonging to the gamble using many distinct but "identically prepared" devices. The expectation is that the totality of such repetitions will yield a frequency of outcomes that tend toward their quantum probability, and this will hold when the process is repeated for the probability assignments to the outcomes of each game in the gamble. Hence, applied *singly* to the outcomes of each game of the gamble, relative frequency plays the same *evidentiary* role in the evaluation of the probability assignments of a quantum gamble as it does in the evaluation of classical probability assignments.

On our proposal for understanding its key conceptual contribution, is there a sense in which the quantum theory is a locally realistic theory? Recall that Einstein's local realism adds to the traditional philosophical idea of realism—the idea that the nature of reality is independent of our capacity to know it—a principle of mutual independence. In Einstein's formulation, once their interaction has ceased and two EPR-correlated systems are sufficiently far apart in space, the value of a parameter of one of the two systems cannot be affected by the kind of measurement performed on the system with which it is paired, nor can it be affected by whether a measurement is or is not performed on the paired system.

Einstein's principle of mutual independence is formulated within the conceptual framework of classical physics, and more generally, within the framework of classical probability theory. But the quantum theory is an "irreducibly statistical theory" that cannot be represented within a classical probabilistic framework. A similar situation arose in connection with completeness. It was natural for Einstein to have supposed that unless the quantum-theoretical description of a physical system includes real factual situations, it cannot be held to be complete. But as we have argued, the conceptual novelty of the theory *consists* in its departure from a probabilistic framework in which probability assignments are always representable as weighted averages of two-valued measures to one in which this is no longer true. In fact, this may be taken as an explication of what is *meant* by the claim that the theory is irreducibly statistical. If therefore we are to have a useful metatheoretical concept of completeness for theories like quantum mechanics, it cannot be uncritically imported from the classical context.

Earlier we argued that there is a concept of completeness that generalizes the classical concept and which was shown by Gleason to apply to the quantum theory. This concept does not require that an irreducibly statistical theory should derive its probability measures from a level of description that corresponds to Einstein's real factual situations. Rather, completeness in the generalized sense established by Gleason requires that the quantum theory should generate all possible positive real-valued measures that are classical probability measures on Boolean subalgebras of the algebra of properties that the theory associates with a physical system. The justification for interpreting the theorem as a result about *probability* assignments—for regarding it as interpretable as a theorem about our applied concept of probability, and not merely a theorem of pure mathematics—is given by Pitowsky's explanation of how the concept of quantum probability agrees with the epistemic understanding of probability in terms of the notion of a fair betting quotient.

Bell's account of a "locally explicable" correlation extracts certain salient features of the causal structure of Minkowski space-time, which then act as a set of constraints on a locally causal explanation of a correlation between spatially separated systems. Bell understood his theorem to suggest that since special relativity is committed to this causal structure *and* to the thesis that the Lorentz group acts transitively in the class of all inertial frames, special relativity is in tension with the prediction of elementary quantum mechanics. By contrast, we have argued that Bell's theorem exposes the failure of the classical theory of probability to account for quantum correlations of the kind exhibited in the

EPR state. The quantum theory calls into question what should be required of the probability theory that replaces the classical theory in order that the new theory might be correctly judged *complete*. The notion of completeness to which we are led is very different from the one deployed by Einstein or EPR, since the appropriate concept of completeness is independent of whether probability measures are representable as averages over real factual situations in the sense of truth-value assignments. Indeed, the quantum theory's irreducibly statistical character *consists* in the fact that its probability measures are *not* in general representable as averages over real factual situations in the sense of truth-value assignments. In the context of special relativity, the irreducibly statistical nature of the implicit probability theory of quantum mechanics is virtually *forced* by the phenomena of quantum correlation and the causal structure of Minkowski space-time. Insofar as special relativity has isolated the true causal structure of space-time and the significance of the universality of the Lorentz group, Bell's theorem shows that the class of appropriate theories of correlations like those exposed by EPR must include one which incorporates among its principles the generalized probability theory implicit in quantum mechanics.[†]

The conclusion we have just reached is related to a discussion of S. Popescu and D. Rohrlich.[49] Following a suggestion of Aharonov, Popescu, and Rohrlich (1994, p. 384) "propose two axioms for quantum theory, nonlocality and relativistic causality, which together imply quantum indeterminacy." Their "nonlocality" is essentially the content of Bell's *theorem* and their "relativistic cau-

---

[†]Editor's note: It turns out that special-relativistic causal structure is not the "true" one, insofar as gravitation turns out to be constrained by *general* relativity—see my Editor's Afterword.

sality" is what I have referred to as Bell's *analysis* of local causality. Together, these axioms tell us that the probabilities of quantum mechanics cannot be captured by the classical concept of probability or, as Popescu and Rohrlich say, they exhibit "quantum indeterminacy." It would be gratifying if one could argue that Bell's theorem shows that the appropriate theory of correlations like those exposed by EPR must be one which incorporates among its principles the generalized probability theory implicit in quantum mechanics, rather than the weaker conclusion that the appropriate theory must be one of a class of theories which includes the generalized probability theory of quantum mechanics among its members. Unfortunately, this stronger conclusion is not justified, for as Popescu and Rohrlich argue,

> From our brief exercise with nonlocal correlations, however, we learn that our two axioms do not determine quantum theory: a theory that allows nonlocal correlations but preserves relativistic causality might not be quantum mechanics but a "superquantum" mechanics. Thus, we have identified a class of theories, to which quantum mechanics belongs, that yield nonlocal correlations while preserving causality (ibid).

If nonlocality and relativistic causality uniquely determined the correlations permitted by quantum mechanics, this would be significant, not so much as a novel axiomatization of the quantum correlations, but as a *deduction from phenomena,* in Newton's sense, of the quantum *theory* of correlated systems. The nonlocal correlations permitted by quantum mechanics are distinguished by the fact that they are among those forced by relativistic causality; but they are *also* distinguished by the fact that they are accounted for by a theory that, despite its irreducibly statistical character, is *complete* in an entirely nonarbitrary sense. Whether the theories of superquantum correlations fall under a comparably convincing concept of completeness is an interesting and, so far, entirely open question.

In order to evaluate whether quantum mechanics is a locally realistic theory, Einstein's mutual independence condition should be generalized to accommodate the theory's irreducibly statistical character: Einstein's formulation refers to the values of parameters, or equivalently, the holding of properties. As in the case of completeness, we require a probabilistic explication of mutual independence, an explication formulated at the surface or operational level of the experimental predictions of the theory, without an appeal to parameters which are invoked to "complete" the quantum-mechanical description in conformity with a classical understanding of completeness. But this is precisely the point of the *no-signaling condition* which tells us that the marginal probability for the directional properties of either of two correlated systems cannot be affected by the *kind* of measurement performed on the system with which it is paired—nor can it be affected by *whether* a measurement is or is not performed on the paired system—once the interaction has ceased. No-signaling is not simply a condition which is arbitrarily imposed on the marginal probabilities of the properties of separated systems: It is arguably the natural probabilistic generalization of Einstein's mutual independence condition for an irreducibly stochastic theory like quantum mechanics. By contrast, the condition of parameter independence *extends* the no-signaling condition from a condition that refers to the surface or operational level of the measurement of directional properties to one that refers to the values of the underlying parameters that are introduced to effect an

Einsteinian completion of the quantum-mechanical description of the correlations. Given its close similarity to Einstein's mutual independence condition, the fact that the theory satisfies no-signaling may be invoked to justify the claim that quantum mechanics is a local theory, albeit a local theory of *nonlocal* correlations, correlations that hold independently of the distance separating the correlated particles.

The question of whether, on our understanding of it, the theory is realistic is subtle. One component of the question of realism is addressed by the question of its completeness. As a consequence of its irreducibly statistical character, the theory is not complete in Einstein's sense, but it is complete in the generalized sense of completeness. It is possible to argue against both our understanding of the completeness of the theory and the idea that our interpretation of it is a realist interpretation on the ground that it fails to preserve the classical spelling out of completeness in terms of Einstein's idea of real factual situations. But this would be to miss what we maintain is the theory's essential conceptual point. On our view, the theory's contribution to our representation of physical reality consists in its identification of a structural principle as a new object of theoretical reflection: Just as special relativity understands the phenomena of dilation and contraction in terms of features of Minkowski space-time, rather than the assumed effect of the electromagnetic character of the constitution of matter, the probabilities exhibited by the EPR correlations are understood as a consequence of the Hilbert space structure of the properties of physical systems, rather than the effect of unknown nonlocal causes. In neither case are the phenomena made any less surprising to classical modes of thinking. But that was never the point of achieving a theoretical understanding of them.

There are certain components of Einstein's realism that are preserved on our understanding of the theory's distinctive conceptual innovation. Despite the limitative results we have discussed, no one can seriously suppose that the interpretational problems of quantum mechanics call into question the existence of atomic and subatomic reality. Similarly, no one can seriously suppose that the theory challenges the fact that elementary particles have various properties independently of our knowledge of them: That an electron is a spin-half particle, that it carries an electric charge and has a rest mass, are just some among the many propositions involving properties of elementary particles which we can unequivocally assert to be true, true independently of any measurement interaction—true even from the perspective of a "detached observer."[50]

The controversy between Einstein and the orthodox quantum theoreticians over quantum mechanics and realism concerns the properties which are the subject of the correlations and no-hidden-variable theorems, properties which are not constant over time, but are associated with dynamical variables that are subject to change. Einstein understood the "orthodox quantum theoreticians" to deny the reality of states of affairs involving such properties in the absence of measurement, so that they might address what he believed EPR had exposed as an incompleteness in the quantum-mechanical description of reality. The present account does not rest on such a denial, since it takes the conceptual point of the theory to consist entirely in what it reveals about probability when it is applied to the values of parameters that are not constant over time. On this account, the quantum-mechanical description of reality is not complete in the sense Einstein required because the probability measures of the theory are not in general representable as averages over two-valued measures; and by the correspondence between two-valued measures and truth-value assignments, neither are they representable as averages over real factual situations—assuming that this idea of Einstein's is adequately explicated by the notion of a truth-value assignment.

While it is certainly possible to extrapolate from facts about how the theory does and does not arrive at its probability assignments to a claim about real factual situations, understanding the conceptual innovation of quantum mechanics does not require such an extrapolation. What we do know on the basis of the limitative and experimental results is that the theory's ability to anticipate phenomena like the EPR correlations is indissolubly bound up with the way the algebraic structure it imputes to physical properties informs its generation of probability measures. Because of the prominence that the quantum theory assigns to the algebra of properties, the theory's contribution to our representation of the structure and constitution of reality is a decidedly structural and principle-theoretic one. In this respect, our proposal for how best to understand the nature of the theory's contribution to our representation of reality is familiar from the case of relativity and the structure of space-time. What is unfamiliar is the idea that the algebraic structure of the possible properties of a physical system should emerge as an object of theoretical reflection and revision.

EDITOR'S AFTERWORD

At the end of my Foreword I connected Demopoulos's discussion of the role of theory-mediated measurements in Perrin's famous argument for molecular reality with the earlier use of such measurements by Newton in his famous argument for the theory of universal gravitation. I emphasized the approximative and iterative character of both arguments, which eventually resulted in revolutionary new theories as replacements for the original (now "classical") theories. Newton's theory of universal gravitation eventually gave way to general relativity in a number of applications in astronomy and astrophysics; classical mechanics eventually gave way to quantum mechanics as the most appropriate description (so far) of molecular, atomic, and subatomic reality. It is by no means surprising, therefore, that Demopoulos frames his concluding discussion in the fourth chapter on quantum reality in terms of Bohr's mature conception of the primacy of classical concepts, which, according to Demopoulos, find their rightful place in the *evidentiary* framework within which the evolving new *theoretical* framework (namely quantum mechanics) is subject to experimental investigation. This conception, in Demopoulos's hands, is thus a natural continuation of the role of theory-mediated measurements in both Newton's argument for universal gravitation and Perrin's argument for molecular reality.

Demopoulos emphasizes, however, that there is an equally important discontinuity in the transition to quantum reality. The latter essentially involves what Bohr called *complementarity*: The idea that a quantum description of reality is always essentially *partial* or *perspectival*, insofar as the theoretical representation of physical quantities can, as a matter of principle, never assign precise values to all such quantities at once (noncommutativity). In the evidentiary framework, moreover, it is similarly impossible, as a matter of principle, that we can precisely measure all such quantities at once (no simultaneously available exact measurements for the totality of quantities). These ideas are characteristic of quantum reality, and the problem, for both Demopoulos and Bohr, is to make sense of them in a clear and precise way. The most important problem for Demopoulos, moreover, is to find nonetheless an equally clear and precise sense of reality in the quantum-mechanical case. I believe that he has indeed found a satisfactory solution to this problem. But his solution has only become completely clear to me by considering the development of his ideas on this subject over time, beginning with his earliest philosophical work in the 1970s and concluding with the present book in 2021.

Demopoulos joined Jeffrey Bub at the University of Western Ontario in the years 1972–1974. He began a fruitful collaboration there with Bub on the logical interpretation of quantum mechanics, inspired by Hilary Putnam's "Is Logic Empirical?" first published in 1968. Putnam there sets up an analogy between the non-Euclidean space-time structure of general relativity and the nonclassical (non-Boolean) algebraic structure fundamental to quantum mechanics. In both cases, Putnam argues that the structure in question is in no way conventional or nonfactual, but rather represents a discovery about the fundamental structure of the physical world. Putnam also argues, however, that quantum mechanics on the logical interpretation is perfectly consistent with the existence of precise values for all physical quantities at once, whether or not they can be simultaneously measured. Bub and Demopoulos basically followed Putnam's example—and also the earlier example of David Finkelstein's "The Logic of Quantum Physics" appearing in 1962–1963—although the Bub-Demopoulos approach is considerably more abstract and rigorous, as well as more thoroughly developed.

The Bub-Demopoulos approach focuses primarily on the algebraic structure of the system of (closed) subspaces of a Hilbert space, whether considered as a nondistributive ortho-normal lattice in the work of Jauch and Piron or as a partial Boolean algebra in that of Kochen and Specker (1967). Bub published *The Interpretation of Quantum Mechanics* in 1974, with a more concise exposition jointly authored by Demopoulos and Bub appearing as a thirty-page article under the same title in the same year. It was Demopoulos alone, however, who provided a rigorous defense of the existence of simultaneous precise values characteristic of the "orthodox" quantum-logical interpretation in "Completeness and Realism in Quantum Mechanics," presented in 1975 and published in 1977. Following the partial Boolean algebra approach of Kochen and Specker, Demopoulos considered a finite number of triples of mutually orthogonal rays in three-dimensional Hilbert space, where each such triple represents the three distinct values of a spin quantity in three-dimensional physical space. We thereby represent a finite system of disjunctive values of spin quantities, where each particular orthogonal triple of rays generates a classical Boolean algebra of subspaces. The entire system of such triples, however, does not generate a Boolean algebra but rather a *partial* Boolean algebra. Nonetheless, each orthogonal triple spans the whole Hilbert space in question and thus yields the value of 1 for each corresponding triadic disjunction, so that the conjunction of all such triadic disjunctions also yields the value of 1. The conjunctive proposition is thus valid (or at least true) in the partial Boolean algebra approach, where each disjunctive proposition then "says" that one and only one of the possible spin values obtains. So, it appears, we have simultaneous precise and determinate values for all such spin quantities.

Demopoulos soon became disenchanted with this solution, however, and he therefore left behind the quantum-logical *interpretation* of the theory for the project of quantum-logical *axiomatization* of the theory in the spirit of the Kochen and Specker partial Boolean algebra approach. The main problem with quantum *logic* to which Demopoulos then repeatedly appeals is that the above quantum-logical proposition is *contradictory* in classical propositional logic. To assert that this proposition is nonetheless *true* in quantum logic, therefore, appears to involve a concept of truth that is well-nigh unintelligible. And, in any case, it is clear that this proposed change of logic is considerably more problematic than the case of intuitionistic logic, which is strictly weaker than classical logic rather than incomparably stronger. It seems that Michael Dummett's work on intuitionistic logic in the theory of meaning and truth contributed to Demopoulos's evolving and changing perspective on the significance of quantum logic. In particular, although Demopoulos refers briefly to Dummett's *Frege: Philosophy of Language* (1973) in "Completeness and Realism in Quantum Mechanics," by the time of his more general discussion of realism and truth in "The Rejection of Truth-Conditional Semantics in Putnam and Dummett" (1982), Demopoulos has considered the full panoply of Dummett's seminal works of this period—from the (first) Frege book (1973), through *The Logical Roots of Metaphysics* (1975, as the William James Lectures), "What Is a Theory of Meaning II" (1976), *Elements of Intuitionism* (1977), and *Truth and Other Enigmas* (1978).

From 1982 to 2004 Demopoulos published little on the interpretation of quantum mechanics. During this period he rather focused on new foundational work in linguistics, computational models of the mind, and learnability—and also on his abiding interest in the history of analytic philosophy. His 1995 book *Frege's Philosophy of Mathematics* (mentioned in my Foreword) is of special significance. Nevertheless, Demopoulos returned to the relationship between classical and quantum mechanics on a couple of occasions—first in a paper on "Elementary Propositions and Independence" co-authored with his colleague John L. Bell (1996) and then in "The Algebraic Basis of Quantum Logic" (2000), appearing as a review of *Quantum Logic in Algebraic Approach* by Miklos Redei.

In 2004, however, Demopoulos outlined a new approach to the interpretation of quantum mechanics in "Elementary Propositions and Essentially Incomplete Knowledge: A Framework for the Interpretation of Quantum Mechanics." Following the 1996 paper with Bell, it begins with the Wittgensteinean idea of the logical independence of all elementary propositions in the *Tractatus*. This idea is now reinterpreted in terms of atomic propositions in a *free* partial Boolean algebra consisting of a family of four-element Boolean algebras comprising only 0, *p*, not-*p*, and 1 for each elementary proposition. A free partial Boolean algebra represents the minimal logical structure that all elementary propositions must have, simply as propositions, and can thus be attributed to them a priori. The point for Demopoulos is that, whereas the full partial Boolean algebraic structure of Kochen and Specker represents the possible distribution of *knowledge* in quantum mechanics given by (generalized) probability measures defined on this structure, the distribution of *truth* over elementary propositions attributing values to fundamental physical quantities involves only the corresponding free partial Boolean algebra. Thus, while quantum mechanics is essentially incomplete with respect to the distribution of knowledge (since there are no generalized probability measures assigning either zero or one to all propositions at once), it can nonetheless remain complete with respect to the distribution of truth and falsity over elementary propositions.

Demopoulos explains this situation by appealing to the contrast between two-dimensional and three-dimensional Hilbert spaces. Two-dimensional Hilbert spaces represent the quantity spin-1/2, which has only two possible values in any given direction in three-dimensional physical space (up or down, left or right, and so on). What corresponds to this in the two-dimensional Hilbert space is a family of orthogonal rays. Each Boolean subalgebra spans the whole space, and no distinct subalgebras (generated by distinct orthogonal rays) have an elementary proposition in common. In the three-dimensional example considered by Kochen and Specker, by contrast, we consider a spin-1 quantity, comprising a family of three orthogonal spin directions (with possible directional values of 1, 0, and –1) in both Hilbert and physical space. Here, however, we have the phenomenon of contextuality, insofar as two distinct Boolean subalgebras can easily have a ray in common in both Hilbert and physical space—consider, for example, a rotation of less than 90° of any three-dimensional orthogonal triple around one of its axes. The elementary proposition corresponding to the fixed axis can have distinct truth values in the context of distinct (rotated) Boolean subalgebras, but it must receive the same probability relative to any such Boolean subalgebra.

The partial Boolean algebra generated in the two-dimensional Hilbert space is degenerate. Although not itself a Boolean algebra, it is still homomorphically embeddable into a Boolean algebra and thus admits a consistent assignment of 1 and 0 (truth or falsity) to all elementary propositions. In three dimensions, by contrast, this is not possible, precisely because of the *non*contextuality of quantum probabilities. This provides Demopoulos with a wedge between truth and knowledge in quantum mechanics, because the noncontextuality of quantum probabilities rests on nothing more nor less than empirical facts concerning the distribution of statistical information in different Boolean subalgebras corresponding to different choices of ideal measuring instruments. The statistical information is invariant under all such choices—as a matter of empirical fact—while the actual measured values are not. The minimal conception of a free partial Boolean algebra, however, involves only the necessary logico-combinatorial features of any proposition as such (0, p, not-p, and 1). And, because comeasurability in general is an empirical rather than logico-combinatorial relation, this is compatible with Demopoulos's conceptual wedge between truth and knowledge in quantum mechanics.

Demopoulos elaborates on this conception in "On the Notion of a Physical Theory of an Incompletely Knowable Domain," published in a volume in honor of Bub in 2006. It provides a more detailed and explicit treatment of three-dimensional spin values in both Hilbert and physical space, together with the corresponding treatment of probabilities. Nevertheless, the implications for quantum-mechanical reality are the same. As far as truth and falsity are concerned, we have determinate values for all physical quantities simultaneously. Our *knowledge* of these values, however, remains incomplete due to the logically necessary absence of two-valued probability measures over the totality of Boolean subalgebras—which, in turn, follows from the general noncontextuality of quantum probabilities.

Demopoulos abandons the divergence between knowledge and truth in "Effects and Propositions," appearing in a special issue of *Foundations of Physics*, again dedicated to Bub, in 2010. Demopoulos does not say so explicitly, but his change of mind perhaps has to do with a kind of vacuity afflicting both the original defense of value determinacy in "Completeness and Realism in Quantum Mechanics" (1977) and the more recent attempt radically to separate the concepts of knowledge and truth (2004–2006). The (1977) paper on completeness allows us to "say" in a quantum-logical proposition that all values of physical quantities are determinate, but it does not make it possible to say anything more specific about the distribution of these values. All that we have, in the end, is a conjunction of quantum-logically valid formulas, each of which is a quantum-logical exclusive disjunction. The radical separation of knowledge and truth in the later period (2004–2006), by contrast, is made possible by a definitive separation of the concepts of elementary proposition and quantum probability. Nevertheless, we can still say nothing specific about the distribution of truth values over all elementary propositions going beyond the quantum probabilities themselves—which, as we know, are (classically) logically inconsistent with simultaneous determinate values. Both of Demopoulos's earlier approaches to quantum determinacy, for this reason, now appear to be purely formal and thus vacuous.

In any case, "Effects and Propositions" represents his definitive rejection of value determinacy as the hallmark of reality in quantum mechanics. The question remains, therefore, of how to understand such reality now. Demopoulos first distinguishes between *eternal* and *dynamical* properties of particles. The former are exemplified by properties like the spin, mass, and charge of the electron. Such properties characterize the kind of particle in question and do not change over time due to dynamical interactions between particles. The dynamical properties of particles—positions, momenta, specific directional spin values, and so on—do change over time, and it is precisely these properties for which we cannot have simultaneous determinate values at any time. Moreover, the fact that dynamical properties are relative to particular measurement contexts seems incompatible with their *objectivity*, insofar as invariance over different measurement contexts (different perspectives) is a criterion of objectivity in general. So we cannot have both determinacy and objectivity for the dynamical properties of particles. The solution Demopoulos proposes is to distinguish, in principle, between the eternal and dynamical properties of particles. The latter are not propositions—properties belonging to particles—at all. They rather characterize the effects of dynamical interactions between the particles themselves and the classically describable instruments with which we measure them. Since any single such instrument in a given context yields only commeasurable effects, we have both objectivity and determinacy for them. And, for the totality of effects, we have objectivity for all probabilities defined over this totality due to the noncontextuality of quantum probability.

The framework of effects, in this way, is close to the approach of Demopoulos's final interpretation in the present book. This becomes clearer in a second paper on the framework of effects appearing two years later, in a volume in honor of the memory of Itamar Pitowsky edited by Yemima Ben-Menahem and Meir Hemmo, entitled *Probability in Physics*. The title of Demopoulos's 2012 paper is "Generalized Probability Measures and the Framework of Effects," and it begins, in the first sentence, with a reference to A. M. Gleason's well-known theorem, "Measures on the Closed Subspaces of a Hilbert Space" (1957), according to which all probabilities employed in quantum mechanics (via either pure or mixed states) are generalized probability measures on these closed subspaces. The generalization in question is from a Boolean algebra (a field of sets in classical probability theory) to the partial Boolean algebra of subspaces of a vector space—which, due to Gleason's theorem, is entirely incompatible (in three-dimensional and higher-dimensional spaces) with two-valued probability measures defined over all such subspaces. As we shall see, the central idea of the present book becomes what Demopoulos will call the "completeness" of quantum probabilities, now taken to be the core of quantum reality.

The view developed in the fourth chapter of this book is a reconstruction of Bohr's mature position, arrived at in his ongoing debate on the status of quantum mechanics with Einstein. The Bohr-Einstein debate culminates in the fundamental paper by Einstein, Podolsky, and Rosen (1935) on the "completeness" of quantum mechanics (EPR), supplemented by Bohr's (and Einstein's) subsequent objections and replies. Since Einstein's conception of *local realism* is Bohr's target here, Demopoulos provides a novel description of J. S. Bell's analysis of this realism—which, in Demopoulos's account, figures essentially in Bell's now-famous theorem on local hidden variables. The point is that Bell's analysis and theorem together provide a strong rebuttal to EPR by showing that Einstein's local realism, so understood, is incompatible with the quantum-mechanical probabilities and thus with quantum mechanics itself. Demopoulos concludes by arguing on this basis for the significance of Gleason's theorem in establishing the necessarily *probabilistic* completeness of quantum mechanics, which thereby provides the theory with its objective empirical reality.

The EPR argument is familiar. We begin with two correlated quantum systems, which, over time, become separated from one another at arbitrary distances. The correlation between the two systems in their joint state is preserved over time, so that, in particular, the correlation between certain eigenvalues of the two systems remains perfect throughout. One can always infer from a measurement of one system the correlated eigenvalue of the other with "certainty" (probability = 1). At this point, however, EPR introduces its "criterion of [physical] reality" (see Chapter 4 in this book): "If, without in any way disturbing a system, we can predict with certainty (i.e., with probability equal to unity) the value of a physical quantity, then there exists an element of physical reality corresponding to this physical quantity. . . . Regarded not as a necessary, but merely as a sufficient, condition of reality, this criterion is in agreement with classical as well as quantum-mechanical ideas of reality." Without begging any questions between classical and quantum conceptions of reality, therefore, it follows that the quantum state cannot be a complete description of quantum reality. This is because the (correlational) joint state that mediates the inference from measurements on one of the two systems to eigenvalues of the other (now spatially separated) system is assumed to remain, which is incompatible with the supposedly resulting eigenstate of the target system. (The latter gives probability $= 1$ for the eigenvalue in question, while the original joint state continues to give probability $= \frac{1}{2}$.)

The EPR argument, understood in this way, is offered as a *reductio* of the completeness of quantum mechanics—one that works for both classical and quantum conceptions of physical reality. But Bell's rejoinder, some thirty years later, is offered, on Demopoulos's account, as a *reductio* of the combination of Einstein's local realism and classical probability theory. What, then, is Einstein's local realism? Here Demopoulos makes a decisive move by taking local realism, for Einstein, as a condition that holds for both Newtonian and relativistic causality—even if we take Newtonian gravitational force to be an immediate action at a distance. Thus, in the penultimate paragraph of Section 4.2, Demopoulos explains that Einstein's local realism belongs to "'the realm of physical ideas independently of the quantum theory' that have their origin in everyday thinking." Demopoulos continues in the same paragraph (emphasis in the original): "Not only is Newtonian mechanics one among the ideas of physics independently of the quantum theory, but the notion that there are quasi-closed systems of the sort that local realism requires is compatible with an action at a distance theory like Newton's *just because that theory allows for the kind of approximative reasoning that the concept of a quasi-closed system rests upon.* Local realism is therefore a constraint on theory construction that is satisfied by Newtonian mechanics and by the conception of theory testing that emerged from the Newtonian tradition of theory-mediated measurement."

This conception of local realism is central to Demopoulos's overall view, and it allows us to connect his final interpretation of quantum mechanics with the Newtonian tradition of theory-mediated measurement—as noted at the beginning of this Afterword and developed at some length in my Foreword. In particular, the discussion in the Foreword supplies ample illustrations of the two key notions in Demopoulos's conception of local realism. In the case of quasi-closed systems, for example, we find that in Newton's account of the solar system the moons of Jupiter and Saturn illustrate this concept extremely well. Although the gravitational field of the Sun is determinative of the orbital motions of both systems as a whole (the systems comprising both the central bodies and their moons), the perturbations of the orbits of the moons around their central bodies due to the Sun are negligible. The action of the gravitational field of the Sun on the two central bodies determines their orbits around the Sun quite well (with the exception of small perturbations of the same two bodies exerted reciprocally by Jupiter and Saturn on one another), but the Sun adds nothing of any consequence to the orbits of these moons around their central bodies. In the case of the Earth-moon system, by contrast, it quickly became clear that the Earth's gravitational field alone is quite insufficient, and it is for precisely this reason that Newton then introduced a perturbation of our moon by the Sun—although he did not succeed in calculating it exactly.

Thus, while the two systems of moons orbiting around Jupiter and Saturn respectively are indeed quasi-closed in the required sense, the same is by no means the case for the Earth-moon system. The difference between these two kinds of examples depends on the much greater relative distances of Jupiter and Saturn from the Sun in comparison with that of the Earth-moon system—such that, in particular, the directions of attraction of the moons of Jupiter and Saturn are thus always very close to being parallel with one another. The Earth-moon system, by contrast, is relatively much closer to the Sun, so that the successive directions of attraction of the moon in its orbit are considerably further from parallelism. Moreover, the much greater distances of Jupiter and Saturn from the Sun, together with the much greater masses of these two central bodies relative to the Earth, implies that the attraction of the relevant moons by their two central bodies (which are of course much closer to them than is the Sun) dominates the Sun's now relatively weak attraction, leaving only a negligible error.

These considerations make it clear that the two concepts of *quasi-closed* system and *approximative* reasoning are indissolubly linked in the Newtonian tradition of theory-mediated measurement. Indeed, as we saw in my Foreword, the approximative reasoning delineated in Rule 4 of the *Principia* (third edition) is essential to the Newtonian method: "*In experimental philosophy propositions gathered from phenomena by induction should be considered either exactly or very nearly true . . . , until yet other phenomena make such propositions either more exact or liable to exceptions.*" And, as we saw in the sequel, this model of "experimental philosophy" is constantly applied in the perturbational analysis of orbital motions throughout the development of Newtonian astronomy, from the three editions of the *Principia*, the following more-refined work of Euler and Clairault, the subsequent analyses of Laplace, the nineteenth-century discoveries of Uranus and Neptune, and finally the discovery of general relativity via extremely small perturbations of the orbit of Mercury. We also saw that, whereas only the last case involved what turned out to be an *exception* to the Newtonian law of universal gravitation (including the more limited inverse-square law), the Newtonian method proposed in Rule 4 continues to remain in full force throughout. In particular, in the absence of this method (and the Newtonian laws that had led us safely up to this point), we would have had no conception of the precise magnitude of the needed relativistic correction eventually provided by Einstein's gravitational theory.

The circumstance that the Newtonian method of theory-mediated measurement outlived the details of the Newtonian theory of universal gravitation demonstrates the difference between theoretical and evidentiary frameworks with particular force and clarity. Indeed, we do not in general want the evidentiary framework to presuppose the corresponding theoretical framework, for this raises obvious doubts about the extent to which the results of the experiments in question (belonging to the evidentiary framework) yield only circular arguments in support of the theoretical framework. So it is all to the good that the Newtonian tradition of theory-mediated measurements clearly and explicitly avoids such doubts. In this sense, as Demopoulos repeatedly emphasizes, a theory-mediated measurement is quite distinct from a typical theoretical *prediction*, understood as a logical consequence of the theory in question.

Demopoulos explains in Section 4.1 that Bohr's doctrine of the primacy of classical (i.e., non-quantum-mechanical) concepts should be taken to characterize the *evidentiary* framework specifically. For this reason, moreover, although Demopoulos takes Bohr to be an insightful analyst of the evidentiary framework, he explicitly criticizes a paper of Camilleri and Schlosshauer—entitled "Niels Bohr as Philosopher of Experiment"—for making it a rational requirement that what happens in the evidentiary framework must ultimately be reducible to the relevant theoretical framework (i.e., quantum mechanics). This is manifestly not true for Newtonian mechanics, and it need not be true for quantum mechanics either. Of course, quantum theory in the broad sense has a much more extensive explanatory scope, as well as considerably more accurate predictions, than Newtonian theory ever had. So quantum theory in this sense is a better candidate for a "theory of everything" than Newtonian theory ever was. Nevertheless, there cannot be a rational requirement that quantum theory—or any other physical theory taken to be fundamental—must actually have such an all-encompassing scope. To think otherwise is to yield to the temptation of confusing a hopeful scientific commitment with an established scientific fact.

More generally, an evidentiary framework is always relatively autonomous from its corresponding theoretical framework in the real world. In particular, an evidentiary framework realized in actual experiments is always afflicted by a necessary lack of perfect exactness at any finite stage of experimental investigation in accordance with Newton's Rule 4. Moreover, there are always contingencies in the design and successful application of an evidentiary framework (via one or another experimental apparatus) relative to any given stage of scientific and technological development. In his crucial Section 4.3, "Bell's Theorem and Einstein's Local Realism," therefore, Demopoulos takes it to be a virtue that Bell's analysis of Einstein's local realism precedes Bell's theorem on hidden variables. What is important, in particular, is that Bell's analysis takes place at the "operational" or "phenomenal" level of experimental results, while avoiding as far as possible the theoretical representation of microscopic dynamical variables in Hilbert space (appealing, for example, to directional *settings* of polarizers instead of directional *properties* of photons). This move represents a profound reversal of Demopoulos's discussion of hidden variables in his earlier work on the interpretation of quantum mechanics, which, unlike his final discussion in this book, always gave precedence to the Kochen and Specker partial Boolean algebra approach over the "operational" or experimental approach of Bell. And, as we have also seen, the corresponding reversal is based on Demopoulos's assimilation of the Newtonian tradition of theory-mediated measurement in the interim.

Bell's operational analysis of Einstein's local realism appeals to the rules of special relativistic causation, according to which the light-cone structure of Minkowski space-time is a constraint on Einsteinian local causality: There can be no causal action of one system on another outside the region common to the past-directed light cones of the two systems. I already said, however, that Demopoulos takes both relativistic and Newtonian causality to satisfy the relevant causal constraints in Bell's *empirical* analysis—but how is this even possible if Newtonian gravitation operates instantaneously? The relevant point here, however, is that we did not know in the seventeenth century, and we do not know the *exact* situation even now, whether Newtonian gravity is really instantaneous or merely propagates very rapidly (as does light and electromagnetic interaction more generally). Moreover, Newtonian gravitational interaction is much weaker than electromagnetic interaction in any case, so that the relative weakness of the former, independently of its speed of propagation, suffices to establish a lack of relevant causal influence for cases in which electromagnetic interaction predominates. Thus, for example, if we consider the first experimental evidence (Roemer in 1762) that irregularities in the orbits of the moons of Jupiter during eclipses (by Jupiter viewed from the perspective of the Earth) are due to the finite speed of light propagation, we find that even the gravitational action of the Sun is quite insufficient to explain these irregularities (which follows from the fact that the system of Jupiter and its moons is quasi-closed relative to specifically gravitational interaction).

Bell is well within his rights, therefore, to take Einsteinian local causality to be uncontroversially relativistic. How, then, does he construct a *reductio* of EPR? He focuses on another pillar of Einstein's overall view: that the statistical nature of quantum theory should, in accordance with (local) realism, follow the example of classical statistical mechanics. Such realism demands, in particular, that every quantum-mechanical probability measure be representable by a weighted average of *two-valued* measures defined over all relevant possibilities. And what Bell's theorem shows (in Itamar Pitowsky's formulation) is that the satisfaction of a Bell inequality by a probability measure is equivalent to that measure being *classical*, that is, as precisely such a weighted average of two-valued measures. According to Bell's theorem, therefore, there can be no classical completion of the quantum probabilities consistent with Einstein's local realism.

The point is more subtle than it first appears, however, because Bell does not appeal directly to Hilbert space as the realization of a classical completion. Indeed, if he did, the Kochen and Specker no-hidden-variable proof would already have answered the question directly by the fact that the relevant Hilbert space representation admits no two-valued probability measures all by itself, quite independently of Einstein's local realism. This is why Bell does not appeal to Hilbert space in either his analysis or his theorem, but rather to a more abstract and less theoretically committed conception of hidden parameters subject to minimal and relatively uncontentious conditions. Bell's contributions thus continue with his method of making only uncontroversial assumptions as far as possible—among which Einsteinean local causality plays a clear and ineliminable role. Only so, from Bell's point of view, can we construct a legitimate *reductio* of EPR.

Demopoulos's novel account of the upshot of Bell's contributions—in both Section 4.3 and the concluding Section 4.4 on "Quantum Mechanics and Reality"—considers all the subtleties in considerable detail, and I urge the reader to study this final section carefully. What appears to be most important here is that there are other types of correlational representations that are stronger than those of quantum mechanics itself—and, in particular, that allow nonlocal correlations while preserving relativistic (local) causality. The Hilbert space representation is one of the theories consistent with Bell's work but by no means the only one. Nevertheless, it is the only such theory that exhibits the *completeness* of its possible probability measures derived from Gleason's theorem. And this kind of completeness, as we saw in my earlier discussion of what Demopoulos called "the framework of effects" in the years 2010–2012, does indeed justify the objective reality of the quantum probabilities.

The objective reality of these probabilities, and their completeness, rest on *noncontextuality*. And the latter, in turn, rests on nothing more nor less than empirical facts concerning the distribution of statistical information in different Boolean subalgebras corresponding to different choices of ideal measuring instruments. A statistical quantum state is invariant under all such choices, while the actual measured values are not. Moreover, as Demopoulos explains, quantum noncontextuality can be seen to embody a natural generalization of relativistic local causality for an irreducibly statistical theory like quantum mechanics (that is, a statistical theory admitting no two-valued measures defined on all events). Quantum mechanics is therefore itself a local theory—although, as Demopoulos himself emphasizes, it is a local theory of nonlocal correlations. In this sense, therefore, the completeness of quantum probabilities deriving from Gleason's theorem also yields a generalized *statistical* version of relativistic local causality.

That the version of local causality under consideration is irreducibly statistical represents the crux of the matter. In Bell's version of the EPR thought experiment, however, we are attempting to infer a value of a directional spin value of one particle from a value obtained by an experimental measurement (via a polarizer) of the other. But the introduction of the measurement apparatus (the polarizer) picks out one particular Boolean subalgebra of commeasurable properties, and the resulting measured values are therefore thoroughly *contextual*—relative, that is, to this Boolean subalgebra. By contrast, the irreducibly statistical state defined on the entire partial Boolean algebra is thoroughly *noncontextual*, so that relativistic local causality in the statistical sense is inconsistent with an assignment of determinate directional properties to the correlated systems.

Thus, once again, the statistical sense of completeness appropriate to the quantum probabilities does not admit a representation of all probability measures as weighted averages of two-valued measures. To be sure, we can and must test the probability measures assigned by the theory via the statistical frequencies of measured values relative to one or another experimental apparatus, and these statistical frequencies must provide either confirmation or disconfirmation for the corresponding theoretical probability measures. If we are lucky, moreover, the resulting statistical frequencies must converge, as it were, toward noncontextuality. But this does not mean, of course, that noncontextuality holds for the individual values measured in the context of one or another experimental apparatus.

As Bohr himself might have said, we must therefore renounce the inference to complete and determinate individual dynamical values of quantum systems—which inference, in general, is incompatible with quantum theory. In return, however, we are rewarded with a complete and noncontextual quantum theory of probability, where the measures in question are themselves both complete and objective as empirical realities. Gleason's deep mathematical theorem on the possible generalized measures on a Hilbert space provides us with a correspondingly deep insight into the radically new algebraic structure of the physical micro-world.

As we have seen, however, there remains an ineliminable divide between the empirical features of the physical world and the purely mathematical structures that are taken to represent this world. Demopoulos's conception here strongly emphasizes the importance of this divide, and that is why he gives priority to Bell's analysis of Einstein's local realism from an explicitly empirical "operational" point of view, as opposed to the purely formal analyses appealing to Hilbert space and the partial Boolean algebras of Kochen and Specker. This shift in viewpoint from the approach of Kochen and Specker to that of Bell signals the most important advance in Demopoulos's intellectual journey. As we have also seen, moreover, this same advance represents his most important engagement with the methodological insights of Newton and his followers in this book. For it is here, in particular, that we find the most subtle and developed discussions of the various cases of approximative reasoning applied to gravitational and electromagnetic interactions (both classical and relativistic), which then turn out to bring us to quantum theory and its distinctive probabilistic structure.

Newtonian empirical methodology, as epitomized in Rules 3 and 4, therefore appears to be still very much alive, even though our fundamental physical theories have continued to change after the formulation of quantum mechanics in the 1920s and, in particular, have introduced more and more complex mathematical structures that could not have been foreseen in advance. This development is not to be deplored but rather to be celebrated, for it leads to open-ended creativity freed up by and embodied in the continuous cumulative interweaving of theory and experiment. Of course we do not and cannot know what may come next in this development, but it turns out to be extremely advantageous in driving the potentially infinite growth of fundamental physics in the real world through an iterative sequence of approximations following Newton's last two methodological rules.

NOTES

Introduction

1. This reevaluation was spearheaded by the work of George Smith. See, for example, Smith (2002) and the references cited therein. My understanding of these developments is indebted to discussions with Robert DiSalle and Michael Friedman.

2. See Millikan (1917, pp. 33–42) for a discussion. The derivation of the relevant functional relation between ne and the ratio of the mobility to the coefficient of diffusion is reviewed in Millikan's Appendix A.

3. In addition to Perrin's experiments based on Brownian motion and carried out within the framework of the kinetic theory of gases, there were theory-mediated measurements of Avogadro's constant based on such diverse considerations as those belonging to electro-chemistry and theory of radiation, as well as much else besides.

4. This observation can also be urged against philosophers of the period who are not identified with the partial interpretation view of theories. For example, Quine evidently believed that our methodological practice is incapable of sharply distinguishing between early corpuscularianism and the molecular hypothesis investigated by Perrin:

> Physical objects [and by extension, molecules and atoms] are conceptually imported . . . as convenient intermediaries—not by definition in terms of experience, but simply as irreducible posits comparable, epistemologically, to the gods of Homer. . . . The myth of physical objects[, atoms, and molecules] is epistemologically superior to most in that it has proved more efficacious than other myths as a device for working a manageable structure into the flux of experience. (Quine 1951, p. 44)

But by the early twentieth century, it became clear that the molecular hypothesis was not "comparable, epistemologically" to the early corpuscularian philosophy, let alone to positing the gods of Homer, and it is important that we understand why.

5. This use of "practical" and "theoretical" derives from Carnap's "Empiricism, Semantics and Ontology," here cited as Carnap (1956b, pp. 205–221).

6. Carnap (1974, p. 256); see also Braithwaite (1953, p. 80) and Nagel (1961, chapter 6).

7. As an example of a view of traditional ontological questions which he rejects, Carnap (1956b, p. 215) cites Bernays (1935), according to which the use of real-number variables for the representation of spatiotemporal coordinates is enough to make a committed Platonist of someone who uses the language of physics. There are different ways of extending the ideas of "Empiricism, Semantics and Ontology" to the realism-instrumentalism controversy. For a discussion of various alternatives, see Demopoulos (2011b), reprinted in Demopoulos (2013).

8. Feigl (1950) is a significant exception to this tendency.

9. Compare Carnap (1956a, p. 45), where the asymmetry is expressed as one that holds between existential hypotheses and statements about the past:

> It may be useful . . . to distinguish two kinds of the meaningful use of "real," *viz.*, the common sense use and the scientific use. Although in actual practice there is no sharp line between these two uses, we may, in view of our partition of the total language L into the two parts L_O and L_T, distinguish between the use of "real" in connection with [the observational part] L_O, and that in connection with [the theoretical part] L_T. We assume that L_O contains only one kind of variable, and that the values of these variables are possible observable events. In this context, the question of reality can be raised only with respect to possible events. The statement that a specified possible observable event, e.g., that of this valley having been a lake in earlier times, is real means the same as the statement that the sentence of L_O which describes this event is true, and therefore means just the same as this sentence itself: "This valley was a lake."
>
> For a question of reality in connection with L_T, the situation is in certain respects more complicated. If the question concerns the reality of an event described in theoretical terms, the situation is not much different from the earlier one: to accept a statement of reality of this kind is the same as to accept the sentence of L_T describing the event. However, a question about the reality of something like electrons in general (in contradistinction to the question about the reality of a cloud of electrons moving here now in a specified

way, which is a question of the former kind) or the electromagnetic field in general is of a different nature. A question of this kind is in itself rather ambiguous. But we can give it a good scientific meaning, e.g., if we agree to understand the acceptance of the reality, say, of the electromagnetic field in the classical sense as the acceptance of a language L_T and in it a term, say 'E,' and a set of postulates T which includes the classical laws of the electromagnetic field (say, the Maxwell equations) as postulates for 'E.' For an observer X to "accept" the postulates of T, means here not simply to take T as an uninterpreted calculus, but to use T together with specified rules of correspondence C for guiding his expectations by deriving predictions about future observable events from observed events with the help of T and C.

10. This applies in equal measure to the conception of the evidentiary framework implicit in the views of some of logical empiricism's severest critics. Thus, for example, Penelope Maddy (2007) locates her disagreement with the program of logical reconstruction in its various accounts of the basis for our acceptance of what she calls "rules of evidence." In the case of the molecular hypothesis, such rules take the form (2007, p. 72): "If such and such results from this Perrin experiment, then there are atoms in the solution." According to Maddy, advocates of logical reconstruction like Carnap take the view that our acceptance of such a rule "is a purely pragmatic matter, a conventional choice of one language over another," by contrast with her own view, according to which "the development of the Einstein-Perrin evidence [is] of a piece with [our] standard methods of inquiry, [requiring] careful examination and justification of the usual sorts" (ibid). But Maddy offers no analysis of what was distinctive about the reasoning Perrin marshaled in favor of the molecular hypothesis except to say that it was based on "careful and meticulous work." And insofar as her notion of a rule of evidence suggests an account of the evidentiary framework within which that hypothesis was evaluated, it is indistinguishable from purely hypothetical reasoning; Maddy's account therefore exhibits many of the same omissions of the program of logical reconstruction that she rejects.

11. It is also what distinguishes the present view from "entity realism," the view defended by Ian Hacking (1983). Entity realism addresses the issue of what Hacking calls "realism-in-general" by exposing the philosophical principle it takes to underpin particular cases of realism. According to entity realism, realism about a special class of entities is justified when we can successfully *manipulate* the entities. As Hacking (1983, p. 22) puts it in an allusion to an extension of the idea behind Millikan's oil droplet experiments: "If you can spray them, then they are real." So stated, entity realism is empty without an account of why our interventions are correctly represented as manipulations of

the entities in question. Entity realism simply takes for granted that an account can be found in our canonical methods of proof and evidence. However, the philosophical problem is to identify the nature and structure of successful applications of these methods so that the relevant existence claims can be recognized as justifiably compelling. This is precisely the issue Hacking's entity realism fails to address.

12. In his semipopular essay "What Is the Theory of Relativity?," published in *The London Times,* November 28, 1919: see Einstein (1919).

13. This is evidently the idea that Einstein (1919) sought to emphasize by contrasting the principle-theoretic character of the special theory with a constructive theory like the kinetic-molecular theory of gases. I therefore disagree with Harvey Brown's (2005) proposal that the theory should be understood as telling us that dilation and contraction depend on the constitution of rods and clocks: "A moving rod contracts, and a moving clock dilates, *because of how it is made up and not because of the nature of its spatio-temporal environment*" (Brown 2005, p. 8; italics in the original). Minkowski space-time is the expression of the dynamical symmetry group of Maxwell's theory because of what the dynamical laws of electromagnetism tell us about *motion,* not because of what, on some possible extension of Maxwell's theory, they might tell us about the constitution of matter.

1. Logical Empiricist and Related Reconstructions of Theoretical Knowledge

1. The partial interpretation account of theories is identical with the view Hempel (1970, p. 146) calls "the standard conception" and Putnam (1962, p. 240) calls "the received view." It can be found in Braithwaite (1953) and Carnap (1956a); it is implicit in Ramsey (1929) and presaged in Schlick (1918).

2. See Carnap (1956a, pp. 41–42, 46–48).

3. In this connection, see Freudenthal (1962) and Hallett (2008).

4. Expounded in Carnap (1963) and extended in various ways to be described below in Carnap (1961). The publication date of Carnap (1963) is a poor guide to the work's date of composition, since the publication of the volume in which it appeared was delayed for many years.

5. By the *matrix* of a Ramsey sentence is meant the formula which results when the "new" existential quantifiers are deleted.

6. "Theories," Ramsey (1929), reprinted in Braithwaite (1931, p. 232). A recursive axiomatization of the L_o-consequences of a first-order theory T in L_o * L_t

by its L_o-sentences is called a *Craig transcription* of T; every first-order theory whose L_o-consequences are recursively enumerable has a Craig transcription. The Ramsey sentence of a theory, like one of its Craig transcriptions, eliminates theoretical vocabulary, but unlike a Craig transcription, it retains the connections between observable properties and relations which are mediated by their association with theoretical properties and relations. This is why the difficulties which Hempel (1965, pp. 214–216) showed to be a necessary feature of the elimination of theoretical vocabulary by reconstructions based on the notion of a Craig transcription are not difficulties for Ramsey-sentence-based reconstructions. For further discussion, see Demopoulos (2011a, section 2) or Demopoulos (2013, chapter 7.2) and Putnam (2012).

7. For an exposition of these matters, see Demopoulos (2007). See also Gupta (2009).

8. Carnap (1961).

9. It is here that the restriction noted earlier regarding the nonuniversality of all the theoretical properties and relations of the theory must be assumed to ensure the distinctness of M and M*.

10. See Gupta (2013, especially section 11) for a critical perspective on "inferential-role" models of meaning.

11. The characterization of the logical empiricist view of theories as syntactic was first advanced by Bas van Fraassen; see especially van Fraassen (1980).

12. See Carnap (1963, p. 963): "I agree with Hempel that the Ramsey sentence does indeed refer to theoretical entities. . . . However, it should be noted that these entities are not unobservable physical objects like atoms, electrons, etc., but rather (at least in the form of the theoretical language which I have chosen in [Carnap 1956a, section VII] purely logico-mathematical entities, e.g. natural numbers, classes of such, classes of classes, etc." This is also the understanding of the theoretical terms belonging to a partially interpreted theory that, in his survey of possible positions regarding existential hypotheses, Feigl (1950) calls "(Va) Formalistic Phenomenalism or Syntactical Positivism." In this paper, Feigl sought to stress the continuity between traditional approaches to the problem of our knowledge of the external world and various views of theories and the nature of theoretical knowledge. As its title suggests, the paper's focus is the status of the existence claims that are central to modern physical theories of the constitution of matter.

13. "Abstract" in the sense that its proof does not exclude the possibility that part of the domain of the model is arithmetical. The main result in van Benthem (1978) is this:

Lemma 3.2. For any L_o-structure M, if M is a model of the L_o-consequences of T, then there exists an $L_o * L_t$-structure N such that N is a model of T and the reduction of N to L_o is an elementary extension of M.

Here L_o and L_t are first-order languages with equality, all of whose vocabulary is, respectively, observational and theoretical. ($L_o * L_t$ is the language generated by the observational and theoretical vocabulary of L_o and L_t.) Note that the lemma assumes that the vocabulary of the theory consists only of theoretical and observational terms. We will soon consider the effect of adding mixed terms.

14. Lewis's response is developed in Lewis (1983, 1984). See also Merrill (1980), which Lewis cites as having influenced his reply. Psillos (1999, pp. 67–68) has argued that Lewis's requirement of uniqueness is an anticipation of his later appeal to natural relations in his reply to Putnam. But it is clear from our earlier discussion that for Lewis, all the heavy lifting to establish unique realizability is done by assumptions which are independent of the requirement that the properties and relations interpreting the O- and T-vocabularies are natural, or that they "carve nature at its joints."

15. Russell (1927); hereafter *AoM*. Newman's objection is presented in Newman (1928). Demopoulos and Friedman (1985) is reprinted in Demopoulos (2013).

16. "What I did realise [on reading your article] was that spacio-temporal [sic] continuity of percepts and non-percepts was so axiomatic to my thought that I failed to notice that my statements appeared to deny it." From Russell's letter to Newman of April 24, 1928, reprinted in volume 2 of his *Autobiography* (1968, pp. 176–177). The reevaluation of Russell's causal theory which follows was prompted by discussions with Anil Gupta.

17. We will see in Chapter 3 that Russell's formulations were equivocal from his earliest writings on this topic.

18. Lewis, citing a preprint of Demopoulos and Friedman (1985), remarks that Putnam's argument was anticipated by Newman. However, it is unclear whether Lewis read Newman's paper, since he mistakenly identifies as "a point of mathematical detail" missing from Newman's presentation the failure to observe the argument's dependence on a nonlogical assumption of cardinality: See Lewis (1984, p. 224, n. 9). But it is abundantly clear from Newman's discussion that this is not a point he missed. The only thing absent from Newman's discussion is the model-theoretic framework which its extension to the partial interpretation view of theories utilizes. Putnam's later modifications of his original argument do not have the same natural connection with Newman's observation, but they pursue the consequences for the theory of reference of the possibility of permuting the domain over which the language is interpreted.

19. In Galavotti (1992), a selection of Ramsey's unpublished writings. Ramsey's discussion here comprises pp. 251–252.

20. Uncharacteristically, Ramsey gives no reference for this remark which suggests that it is based on personal communication with Russell. Discussions with Ramsey are acknowledged in the Preface to *AoM*.

21. Russell's construction is discussed in Demopoulos (2003b), and at greater length in Bostock (2012).

22. See Campbell (1920). For a concise account of Campbell's views on theories, see Buchdahl (1964).

23. *AoM*, p. 7, cited by Ramsey in Galavotti (1992, p. 251).

24. Galavotti (1992, p. 251) with a trivial change of class-abstract notation. For a discussion of Eddington which places him in the context of issues raised by Russell's structuralism, see Solomon (1989). I have little to contribute to the discussion of Eddington except to say that although I agree with Solomon that Eddington was attracted to the idea that fundamental physical principles are a priori, I find it implausible that he would have regarded an argument like Newman's as a suitable basis for such a claim.

25. Ramsey's views on *Principia* are discussed in Demopoulos (2013, chapter 11) as is his notion of tautologousness, which is much wider than truth-table decidability.

26. Galavotti (1992, p. 252). I have corrected Ramsey's page reference to *AoM* ("ps 90 ff"), which is obviously incorrect.

27. This objection was first raised in Demopoulos (2003a), reprinted as chapter 6 in Demopoulos (2013); it is also discussed in chapter 4 of that book. The present discussion refines and clarifies these earlier presentations.

28. See Section 2.4 in the next chapter.

29. Especially by Putnam (1962) in his celebrated essay "What Theories Are Not."

30. My use of "phenomena" is nontechnical and follows standard physical practice, which allows for subatomic, and hence unobservable, phenomena. This involves a slight departure from van Fraassen, for whom "phenomena" is sometimes restricted to observable phenomena, as in his assertion that the aim of science is to save the phenomena—a claim that is clearly intended to be interchangeable with the assertion that its aim is to save the appearances.

31. Van Fraassen (1980, p. 64).

32. That this is the main point of van Fraassen's use of the semantic view of theories has been missed by Halvorson (2012, 2013), who argues against constructive empiricism's endorsement of the semantic conception by appealing to the utility of the logical notion of a theory as a set of sentences in a formalized

language. But van Fraassen can certainly concede the utility of the logical notion while maintaining the superiority of constructive empiricism's explication of empirical adequacy over one given in terms of vocabulary. For a response to other aspects of Halvorson's criticisms of the semantic view of theories, see Glymour (2013).

33. This contrast between Carnap's Ramsey-sentence reconstruction and constructive empiricism was first observed by Friedman (2012).

34. It should be noted that there are two inequivalent ways of developing the model-theoretic argument against the partial interpretation reconstruction. An empirically adequate theory might be capable of being shown to be true, because the argument establishes that it is true in an *expansion* of the intended model of the true observation sentences; or the argument might only show that it is true in an *extension* of this model. It can be plausibly argued that Carnap's neutralism about the realism-instrumentalism controversy requires that a partially interpreted theory should be true in an expansion—rather than in an extension—of the model of the true observation statements; otherwise it is not clear that the disagreement between realists and instrumentalists is correctly represented as a disagreement over the acceptance of theoretical vocabulary. For a discussion of this point, see Demopoulos (2013, chapter 4, section 3).

2. Molecular Reality

1. My formulation of the molecular hypothesis basically follows Maxwell (1875, pp. 36–49).

2. This classification of the propositions of the molecular-kinetic theory into constructive and principle-theoretic components is not intended to be exhaustive. Among other things, it does not cover the many propositions which express functional dependencies among the thermodynamic and other parameters of the theory; nor does it attempt to offer a useful classification of them.

3. See van Fraassen (2009) for a forceful presentation of this view.

4. This is argued in detail in G. Smith and R. Seth (2020), *Brownian Motion and Molecular Reality: A Study in Theory-Mediated Measurement.*

5. For example, by contrast with Perrin, J. Loschmidt's estimate of Avogadro's number rested on the assumption that when liquefied, the molecules of a gas are closely packed spheres. See Perrin (1910, section 11). A related point is emphasized in his Nobel Lecture (Perrin 1926, p. 10), quoted later in this chapter.

The issue is raised in Chalmers (2011, pp. 713–715) and discussed in some detail by Smith and Seth (2020).

6. See Einstein (1905, pp. 1–2). As noted in Renn (2005, p. 25), Boltzmann held that such a possibility was excluded by a kind of limit argument, which would imply that there could not be a molecular-kinetic-theoretical explanation of Brownian motion (Boltzmann 1964, p. 318):

> In the molecular theory we assume that the laws of the phenomena found in nature do not essentially deviate from the limits that they would approach in the case of an infinite number of infinitesimally small molecules. . . . It is indispensable for any application of the infinitesimal calculus to molecular theory; indeed, without it, our model which strictly deals always with a large finite number, would not be applicable to apparently continuous quantities. This assumption will seem best justified to those who have carefully considered experiments for the direct proof of the atomic constitution of matter. Even in the smallest neighborhood of the tiniest particles suspended in a gas, the number of molecules is already so large that it seems futile to hope for any observable deviation, even in a very small time, from the limits that the phenomena would approach in the case of an infinite number of molecules.

I am indebted to Melanie Frappier for bringing Renn's paper to my attention.

7. The term "molecular diameter" may seem to imply a commitment to the shape of molecules. However, this is not how the term was always used. As Maxwell expresses it, "the molecular diameter" refers to the distance between the centers of mass of two molecules when they act on one another so as to have "an encounter." The volume of a sphere having this diameter is what is meant by "the volume of the molecule," etc. See Maxwell's (1875, p. 41). This is also true of Perrin's usage of the term "molecular volume": see Perrin (1916, p. 77).

8. Von Smoluchowski (1906, p. 762), explains the fallacy in Nägeli's argument; his discussion is reviewed in von Plato (1994, p. 130). Perrin's long quotation from Father Carbonnelle (Perrin 1910, p. 4) addresses the same objection, noting that the surface of the suspended particles must be below a certain minimum in order that the molecular collisions cease to balance one another, but I have been unable to verify whether Carbonnelle was specifically addressing Nägeli.

9. What is measured and found to be random is the direction of the particle's displacement in the horizontal plane, the plane orthogonal to the direction of gravity: Provided the time interval is not too small, the displacement is doubled when the duration of the displacement is quadrupled, increased tenfold when the duration is increased a hundredfold, and so on.

10. Perrin (1910, sections 13 and 28).

11. See Cohen and Schofield (1978, p. 106).

12. Perrin (1916, pp. 206–207).

13. These are summarized by Perrin (1910, section 42) and Perrin (1916, section 120).

14. Compare Norton (2000, p. 73): "The agreement of all these different methods for estimating N is to be expected if matter has atomic constitution. If, however, matter were not to have atomic constitution, then it would be very improbable that all these estimates of a nonexistent quantity would turn out to agree."

15. Perrin (1910, p. 46).

16. Psillos (2011, p. 17). I have substituted 'N''' for 'N' to accord with the text of Perrin's discussion of the vertical concentration experiments.

17. Not all probabilistic reconstructions take this path. As noted by van Fraassen (2009, p. 6, n. 7), the probabilistic reconstruction set forth in Achinstein (2001) does not attribute such a strategy to Perrin but claims only to represent the fact that the molecular hypothesis has a probability strictly greater than one-half. This may indeed be all that is needed to capture Perrin's remark that the kinetic theory's correct prediction of N' cannot be attributed to mere chance. But it is far too weak to capture the epistemic weight Perrin's investigations were accorded.

18. The difficulty associated with the elision of the distinction between the elimination of alternatives and the elimination of known alternatives is discussed at length in Stanford (2009). While I agree with Stanford on the importance of the distinction, the skeptical conclusions regarding the molecular hypothesis that he draws from his discussion seem to me wholly unwarranted.

19. This is not a claim about Perrin's intentions, but a claim about what was ultimately established by the scientific developments he initiated. Nevertheless, it seems undeniable that the claim correctly captures what Perrin understood to be the main upshot of his contributions to the investigation of Brownian motion. I therefore disagree with van Fraassen (2009), who interprets Perrin as having only been concerned to show the empirical determinability of various parameters of the molecular-kinetic theory, so that the theory could be seen to be "empirically grounded." I do however fully endorse van Fraassen's remarks, directed against Maddy and others, regarding the difficulty with maintaining that the molecular-kinetic theory was even empirically adequate, let alone true.

20. I am indebted to Michael Friedman for bringing this passage of van Fraassen's to my attention.

21. Perrin (1916, p. 89; emphasis added). This remark is followed by an exposition and discussion of his vertical concentration experiments with granules.

22. Gouy's contributions are summarized by Perrin (1910, pp. 4–6; and 1916, pp. 83–85).

23. I am indebted to Robert DiSalle for discussions regarding Perrin's argument in light of the distinction between principle and constructive components of theories.

24. In particular M. von Smoluchowski (1906); for a discussion of von Smoluchowski and a comparison of his investigations with those of Einstein, see von Plato (1994, chapter 3.4).

25. See Poincaré (1912a); p. 224 of the translation in Demopoulos, Frappier, and Bub (2012). Poincaré's views on this and related methodological issues are discussed in greater detail in Chapter 3.

26. Perrin's analogy is drawn in *Atoms* (Perrin 1916, p. 106).

27. See van Fraassen (2009, section 3.2). Van Fraassen remarks on the fact that his notion of groundedness is a development of Clark Glymour's idea of bootstrapping, which is the central idea of *Theory and Evidence* (Glymour 1980).

28. Reichenbach (1920, chapter 4) was concerned with the problem of formulating a notion of truth for systems of equations expressing functional dependencies among various parameters without appealing to their correspondence with reality. He sought to explain the truth of such systems of functional dependencies in terms of *uniqueness of coordination*: a system of equations is true if it is concordant in the values it predicts for the parameters with which it deals; i.e., for any parameter of interest, the system of functional dependencies determines a single (and hence, "uniquely coordinated") value for it (Reichenbach 1920, p. 43). In this respect, Reichenbach's views anticipate van Fraassen, who cites with approval the focus on unique coordination when expounding the notion of empirical grounding: See van Fraassen (2009, p. 12). It is worth remarking that Perrin's interpretation of the significance of the concordance of various determinations of Avogadro's number would have been entirely missed had he followed Reichenbach, for whom the mere fact of concordance suffices as a complete account of the truth of a system of functional dependencies.

29. See Perrin (1910, p. 7).

30. Thomson himself avoids the use of "electron" for the carrier of negative electricity, and he adheres to the traditional terminology established by G. Johnstone Stoney, according to which the term refers to the natural unit of electricity. Considerations in favor of this terminology are given in Millikan (1917, pp. 25–27). Smith (2001, p. 57) suggests that Thomson also wished to distinguish his carriers of negative electricity from the electron theory of Larmor.

31. The pairing of Thomson's and Perrin's contributions to the vindication of atomism goes as far back as the fourth edition of Ostwald's textbook (1909), where they are together singled out for special emphasis when Ostwald withdraws his earlier characterization of the molecular hypothesis as a mere convenience.

32. Thomson's Nobel Lecture is an elegant reconstruction of his contributions and those of his Cavendish collaborators.

33. Norton (2000, pp. 76–77) claims that Thomson proposes to derive the corpuscular nature of cathode rays from the mere concordance of values of e/m, and he then proceeds to criticize him for this. But this is not the structure of Thomson's argument in his Nobel Lecture. Nor does he argue this way in Thomson (1897).

34. This is not the currently accepted figure (which I believe is closer to 1835), but what is important is that it is the correct order of magnitude.

35. Summarized in his Nobel Lecture (Millikan 1924), "The Electron and the Light-Quant from the Experimental Point of View," and discussed at greater length in Millikan (1917, especially chapters 3 and 4).

3. Poincaré on the Theories of Modern Physics

1. Normally waves of lower frequency advance through a medium with greater speed than waves of higher frequency because of the relation between the velocity of the waves and the refractive index of the medium. When a medium exhibits anomalous dispersion, the speeds of, for example, red and violet light are reversed.

2. The quotation is from the English translation of the first (1893) edition of Nernst's textbook, quoted by Smith and Seth (2020, p. 64), to whom my discussion is indebted; see also van 't Hoff's (1901) Nobel Lecture.

3. Compare van 't Hoff (1901).

4. Chapter 9 of *Science and Hypothesis* (Poincaré 1902).

5. Referring to the atomic hypothesis, Poincaré writes: "Hypotheses of this sort have therefore only a metaphorical sense. The scientist should no more interdict them than the poet does metaphors; but he ought to know what they are worth. They may be useful to give a certain satisfaction to the mind, and they will not be injurious provided they are only indifferent hypotheses" (Poincaré 1902, p. 142 of the Halsted translation).

6. "Continuity" is a somewhat inadequate term for expressing Poincaré's idea: It is not merely the use of continuous mathematics in physics that is at issue,

but the nature of the underlying medium as well. But it would be incorrect to express Poincaré's thought with "the continuum" in place of "continuity," given the constructive nature of his view of infinity. As noted in Demopoulos, Frappier, and Bub (2012, p. 223), Poincaré uses a neologism which, as Melanie Frappier has observed, translates virtually directly as *continuism;* this seems intended to address these considerations, and it contrasts simply with its contrary, atomism. By contrast with the preoccupation of contemporary analytic metaphysicians with the existence of a "fundamental level," Poincaré's interest in the continuism-atomism debate is motivated by the place of continuity, in all its manifestations, in current and future physics.

7. But see the final paragraph of the introduction in Demopoulos, Frappier, and Bub (2012), which indicates how the question has acquired a subtlety in contemporary physics that Poincaré could not have envisaged.

8. Poincaré (1912b, p. 90).

9. See, e.g., Krips (1986, pp. 46–47), who argues that "[Poincaré's] reason for accepting the existence of atoms can be seen as a special case of what Whewell called 'consilience of inductions,' i.e., an hypothesis gains support because it figures essentially as part of several different explanations of . . . the *same* phenomenon." Krips's exclusive focus on consilience of inductions is encouraged by the unauthorized translation of the passage from *Science and Hypothesis* that we remarked on earlier. The passage is quoted by Krips (1986, p. 58) and plays an important role in his account. It will be recalled that the unauthorized translation fails to capture Poincaré's separation of mere predictive success from the discovery of true relations.

10. In the text surrounding our quotations from his (1912a) and (1912b), Poincaré recognizes the distinctive confirmation value of independent sources of evidence for the molecular hypothesis and to this extent showed his appreciation of the importance of considerations deriving from consilience.

11. Laudan (1981, pp. 168–169; all italics Laudan's). To his quotation from Whewell's *Philosophy of the Inductive Sciences* (vol. 2, p. 65), Laudan adds a footnote, in which he remarks that "[i]n a similar vein [Whewell] observes [on p. 285 of the same work] that 'when the explanation of two kinds of phenomena, distinct and not apparently connected, leads us to the same cause, such a coincidence does give a reality to the cause, which it has not while it merely accounts for those appearances which suggested the hypothesis.'"

12. Laudan (1981, p. 169, Laudan's italics).

13. Quoted by Worrall (2012, p. 81). Worrall's quotation is from the Dover edition of *Science and Hypothesis,* pp. 149–150; for the authorized Halsted edition,

see p. 133. There is no essential difference between the translations of this passage in the two English editions.

14. Quoted by Laudan (1981, p. 209).

15. Russell (1905, pp. 76–77, Russell's italics). Russell's review appeared in 1905 in *Mind*. I am indebted to Stathis Psillos for reminding me of Russell's review and for calling my attention to the fact that it contains what is likely the earliest formulation of his structuralism.

16. The remark Russell quotes from Poincaré's Preface is directed at Édouard le Roy.

17. Russell's statement is ambiguous. When he writes "the relations are for the most part unknown," does he mean that most relations between material objects are unknown? Or does he mean that every such relation is for the most part unknown—that is, unknown aside from its mathematical properties? I take him to be claiming the latter, but nothing in my discussion hinges on this choice, since I intend to show that Poincaré's views about relations are orthogonal to both claims. It is interesting to note that this ambiguity appears at a very early stage of Russell's thinking about these issues and, as we noted earlier (in Section 1.5), it persists in his exposition of his views as late as *AoM*.

18. Poincaré (1906, p. 142). Translation from the French by Melanie Frappier. I am indebted to Stathis Psillos for calling my attention to Poincaré's letter.

4. Quantum Reality

1. Bohr (1927 / 1961, p. 53). I have silently deleted the definite articles that are grammatically unnatural in English from the phrases, "the classical physical ideas" and "the classical concepts."

2. The general philosophical point which underlies this observation has been given a very accessible formulation and defense in Gupta (2011, pp. 164–195). The observation is of course compatible with the possibility that the presuppositions on which the evidentiary framework rests approximate those of a preferred framework whose principles we believe to be true. We will return to this point and consider in some detail the relation between the conceptual frameworks of classical and quantum mechanics.

3. The 1931 Springer volume to which Schrödinger refers was published by Cambridge University Press in 1934, in English translation, under the title *Atomic Theory and the Description of Nature*. I have inserted in parentheses after Schrödinger's page numbers the pages where these quotations appear in the 1961 Cambridge edition.

Editor's note: There are two passages cited by Schrödinger not occurring above. The first comes from Bohr (1935a, p. 701): "the indispensable use of classical concepts in the interpretation of all proper measurements." The second (and last) passage is also connected to Bohr, and, indeed, it clearly once had a place in Bohr (1935a). One can see this from Bohr's (1935b) answer to the "Einstein paper" (i.e., EPR), where he [Bohr] says that he has revised it for the final version of Bohr (1935a). Thus Bohr (1935b / 1996, 511) [in the seventh volume of Bohr's complete works]: "As far as my answer to the Einstein paper is concerned, I believe that it has already appeared, and you will see that on several points I have tried to express myself more clearly, and I hope that also your first objection concerning the measuring device has thereby been answered. Furthermore, I have left out the reference to the possible significance of the atomic constitution of all measuring instruments for the solution of the still unexplained difficulties in electron theory. The reason is that together with Rosenfeld I am just about to finish a paper about the measuring problems in electron theory in which this question will be elucidated somewhat more fully[35]." One should compare "the consideration of the atomic constitution of all measuring instruments" (Schrödinger letter) with "the possible significance of the atomic constitution of all measuring instruments for the solution of the still unexplained difficulties in electron theory." And, finally, if one follows up footnote 35 at the end of the quote from (1935b / 1996, 511) above, one arrives at the Introduction by Kalckar at the beginning of the seventh volume and sees quite clearly that the problem bothering Bohr and Rosenfeld, together with Heisenberg and Pauli, involved an attempt to shed light on quantum electrodynamics beginning in the early 1930s and extending into the 1950s.

4. Bohr (1958 / 1996, p. 389).

5. See Friedman (2017, p. 209). He there notes that "the 1919 observatory at Sobral in Brazil was less than 10 cubic meters in volume; the stars observed belonged to the Hyades cluster, the center of which is approximately 150 light-years from the earth. In general, in any Riemannian manifold of any curvature and dimension, geometry in the infinitesimally small (of the tangent space at any point) is nonetheless flat or Euclidean."

6. See Malament (1986), which refers to Ehlers (1981).

7. Camilleri and Schlosshauer (2015, p. 74). The citation of Bohr is from (1935b, p. 701). Thanks to Jeffrey Bub for calling my attention to the Camilleri and Schlosshauer paper and to the discussion of Landsman considered below.

8. See Landsman's 2018 reprint (http://arxiv.org/pdf/0804.4849.pdf pp. 2–3); all italics are Landsman's.

9. Compare the Harrigan and Spekkens (2010, pp. 125–157) discussion of *ontic* and *epistemic* interpretations of the wave function.

10. Shimony (1993, p. 89) reports a 1938 discussion between Peter G. Bergmann, Valentine Bargman, and Einstein "during which Einstein took von Neumann's book from the shelf and pointed to premise B_2 of von Neumann's theorem (in section 1 of chapter IV): 'If R, S, . . . are arbitrary quantities and a, b, . . . real numbers, then $\text{Exp}(aR + bS + . . .) = a\text{ExpI} + b\text{Exp}(S) +. . . .$' Einstein then said that there is no reason why this premise should hold in a state not acknowledged by quantum mechanics if R, S, etc. are not simultaneously measurable."

11. See especially the following passage (Einstein 1949, pp. 671–672; italics added):

> The attempt to conceive the quantum-theoretical description as the complete description of the individual systems leads to unnatural theoretical interpretations, which become immediately unnecessary if one accepts the interpretation that the description refers to ensembles of systems and not to individual systems. . . . There exists, however, a simple psychological reason for the fact that this most nearly obvious interpretation is being shunned. For if the statistical quantum theory does not pretend to describe the individual system (and its development in time) completely, it appears unavoidable to look elsewhere for a complete description of the individual system; in doing so it would be clear from the very beginning that the elements of such a description are not contained within the conceptual scheme of the statistical quantum theory. With this one would admit that, in principle, this scheme could not serve as the basis of theoretical physics. *Assuming the success of efforts to accomplish a complete physical description, the statistical quantum theory would, within the framework of future physics, take an approximately analogous position to [that of] statistical mechanics within the framework of classical mechanics.* I am rather firmly convinced that the development of theoretical physics will be of this type; but the path will be lengthy and difficult.

12. The view to the contrary has been a constant source of confusion in the literature about hidden-variable alternatives to quantum mechanics. Kochen and Specker (1967) misrepresent the significance of their investigations and of Bohm's contribution when they claim that their results exclude a theory like Bohm's. See the concluding paragraph of section 4 of their paper.

13. See Einstein's letter (May 12, 1952), in Born (1971d, p. 193).

14. See Myrvold (2003, section 3.1) for an account of early objections to Bohm's theory, including the one that Einstein advanced in his contribution to Born's *Festschrift*. Einstein's objection turns on the fact that Bohm's theory vio-

lates "the well-founded requirement, that in the case of a macro-system the motion should agree approximately with the motion following from classical mechanics" (Einstein 1953, p. 39), translated and quoted by Myrvold, p. 10; or as Bohm (1953, p. 16) expresses the condition: "microscopic theories must always become identical with previously accepted macroscopic theories, when one considers sufficiently large dimensions."

15. See Einstein, Podolsky, and Rosen (1935, pp. 777–780).

16. See Einstein (1935). Yemima Ben-Menahem has emphasized how this letter makes clear Einstein's agreement with the basic argument of the EPR paper and the importance of the conclusion about completeness it sought to establish. This runs counter to claims by Don Howard and Arthur Fine that Einstein disagreed fundamentally with the EPR paper. The main piece of historical evidence cited in support of such claims is Einstein's admission, in a June 1935 letter to Schrödinger, that the paper was written by Podolsky. But aside from what Einstein regarded as the paper's unnecessarily distracting technical presentation, his major dissatisfaction with Podolsky appears to have had little to do with the content of the paper, but to have stemmed from an interview Podolsky gave to the popular press prior to the paper's publication. On Einstein's view of the EPR argument, see Ben-Menahem (2015), and her (2017, pp. 80–88, n. 12). For the episode involving Podolsky, and Einstein's reaction, see Bub (2016, p. 44).

17. Born (1971c, p. 189).

18. Einstein (1935, p. 458; emphasis added).

19. EPR's criterion of reality:

> If, without in any way disturbing a system, we can predict with certainty (i.e., with probability equal to unity) the value of a physical quantity, then there exists an element of physical reality corresponding to this physical quantity. It seems to us that this criterion, while far from exhausting all possible ways of recognizing a physical reality, at least provides us with one such way, whenever the conditions set down in it occur. Regarded not as a necessary, but merely as a sufficient, condition of reality, this criterion is in agreement with classical as well as quantum-mechanical ideas of reality. (*EPR*, pp. 777–778)

20. Bohr (1958, pp. 312–313); reprinted in (1996, pp. 393–394; emphasis added).

21. See Bohr (1935b, p. 700): "Of course there is in a case like that just considered no question of a mechanical disturbance of the system under investigation during the last critical stage of the measuring procedure."

22. Einstein (1948); reprinted with an English translation in Born (1971b, pp. 168–173).

23. For example, it lies at the center of Michael Dummett's many discussions of realism and anti-realism; see especially Dummett (1987).

24. The argument that is spelled out in what Einstein in his letter to Popper explicitly represents as the main idea of the EPR paper concludes that the possibility of assigning more than one ψ-function to a system, on the basis of a measurement performed on the system with which it is paired, is incompatible with the ψ-function being a complete description of the system's state of reality. Howard (1989, pp. 232–233) claims that this argument is unique to Einstein and without parallel in EPR; but the argument is alluded on to on p. 779 of the EPR paper.

25. I have used the translation of Howard (1985, p. 185) because it is more natural than the one given by Born (1971b, p. 171), who translates Einstein's "Prinzip der Nahewirkung" slightly archaically as the "principle of contiguity." ("Unmittelbaren," which Howard translates as "immediate," is italicized by Einstein.)

26. Quoted and translated in Howard (1985, p. 188).

27. Ibid.

28. Quoted and translated in Howard (1985, p. 187).

29. See Einstein's letter to Born of March 3, 1947; reprinted in Born (1971a, pp. 157–158).

30. Bell (1964). For an illuminating discussion of why EPR's argument is inconclusive for purely logical reasons, and therefore, independently of developments stemming from Bell's theorem, see Stairs (2011).

31. Pitowsky's characterization was expressed in correspondence. I recorded it once before, in section 4 of Demopoulos (2010, pp. 368–389).

32. Condition (2) addresses Einstein's criticism of von Neumann's analysis of the problem of hidden variables, which is noted in footnote 10 above.

33. Pitowsky (1994).

34. This is a consequence of the Weyl-Minkowski theorem concerning the dual representation of convex polytopes in terms of their facets or their vertices and the representation of correlations by the notion of a correlation polytope. These concepts are explored in Pitowsky (1989, chapter 2).

35. The exposition that follows is based on Bell (2004).

36. Bell (2004, p. 243). I have omitted some of the equation numbers.

37. The notion of free choice appealed to here is uncontentious and standard in the literature. For a discussion, see Norsen (2009); see also the discussion in Bell, Shimony, Horne, and Clauser (1985). Norsen (2009, pp. 283–284) contains a passage quoted from Bell, et. al. (1985).

38. Bell (2004, p. 243).

39. The duality of these representations applies even when the correlation polytope is not a simplex; however, when it is a simplex, the representation in terms of two-valued measures is unique. I am indebted to Jeffrey Bub for this observation.

40. Tim Maudln has long argued that no classical completeness assumption—indeed, no assumption of "classicality"—informs Bell's argument against EPR. As an observation about what we have isolated as Bell's *analysis,* this is correct. But Maudlin misses the connection with classical completeness that is implicit in Bell's *theorem* and therefore overlooks the role of classicality in Bell's overall argument, which consists of both his analysis and theorem. See for example Maudlin's contribution to his controversy with Reinhard Werner: https.//arxiv .org/abs/1408.1828 and https.//arxiv.org/abs/1411.2120.

41. This characterization of the connection between parameter independence and no signaling is emphasized by Bub (2016, pp. 76–77).

42. See the "toy example" of a stochastic theory of "quantum-like" correlations in Myrvold (2016, section 4.1); sections 4.2 and 4.3 of his paper discuss its possible significance.

43. See the extended passage from Einstein's *Reply* quoted above in Section 4.2, note 11..

44. The sense in which quantum probabilities are manifested by relative frequencies in the probabilistic predictions of particular measurement results is subtle. For a general discussion of some of the methodological issues, see section 3 of Pitowsky (1994). We will return to the bearing of relative frequencies on the confirmation of the probability assignments of the quantum theory in Section 4.4.

45. Bohr (1935b, p. 700; italics Bohr's).

46. See notes 12 and 13 above.

47. See Michel Janssen (2002, p. 499):

According to [Lorentz's] electromagnetic program, the common origin underlying universal Lorentz invariance is that all matter is made of electromagnetic fields and is thus governed by Maxwell's equations. Since Maxwell's equations are Lorentz invariant, all systems in motion must contract. The program initially showed promise, but did not pan out.

48. It is however not without its critics; as I noted in the Introduction, it has been challenged by Harvey Brown (2005): compare note 13 in the Introduction.

49. Popescu and Rorlich (1994). I am indebted to Jeffrey Bub for calling my attention to this paper.

50. The expression is Wolfgang Pauli's. See his letter to Born of March 3, 1954, in Born (1971, pp. 217–218). See also the important and insufficiently cited discussion of the respects in which realism is preserved in quantum mechanics by Yemima Ben-Menahem (1988).

BIBLIOGRAPHY

Achinstein, P. (2001). *The Book of Evidence* (New York: Oxford University Press).

Bell, J. S. (1964). "On the Einstein-Podolsky Rosen Paradox," *Physics* **1**: 195–200.

———. (2004). "La Nouvelle Cuisine," in *Speakable and Unspeakable in Quantum Mechanics* (Cambridge: Cambridge University Press): 232–248.

Bell, J. S., A. Shimony, M. A. Horne, and J. F. Clauser. (1985). "An Exchange on Local Beables," *Dialectica* **39**(2): 85–110.

Ben-Menahem, Y. (1988). "Realism and Quantum Mechanics," in A. van der Merve, F. Selleri, and G. Tarozzi (eds.), *Microphysical Reality and Quantum Formalism* (Dordrecht: Kluwer): 103–113.

———. (2015). "Nonlocality and the Epistemic Interpretation of Quantum Mechanics." http://philsci-archive.pitt.edu/11568/.

———. (2017). "The PBR Theorem: Whose Side Is It On?" *Studies in History and Philosophy of Modern Physics* **57**: 80–88.

Bernays, P. (1935). "Sur le platonisme dans les mathématiques," *L'Enseignement Mathématique* **34**: 52–69, translated by C. Parsons as "On Platonism in Mathematics," and reprinted in H. Putnam and P. Benacerraf (eds.), *Philosophy of Mathematics* (Cambridge: Cambridge University Press, 1983): 258–271.

Bohm, D. (1953). "A Discussion of Certain Remarks by Einstein on Born's Probability Interpretation of the ψ-function," in E. Appleton (ed.), *Scientific Papers Presented to Max Born* (New York: Hafner): 13–19.

Bohr, N. (1927). "The Quantum Postulate and the Recent Development of Atomic Theory," in *Atomic Theory and the Description of Nature* (Cambridge: Cambridge University Press, 1961): 52–91. *Atomic Theory and the Description of Nature* was first completely assembled with four chapters and an Introductory Survey by Springer in German (1931); the first complete English edition was published by Cambridge in 1934 and then reprinted in 1961.

———. (1935a). "Can Quantum Mechanical Description of Physical Reality Be Considered Complete?" *Physical Review* **48**: 696–702.

———. (1935b). "Letter to Schrödinger of October 26th, 1935," in J. Kalckar (ed.), *Niels Bohr: Collected Works. Volume 7: Foundations of Quantum Physics II (1933–1958)* (Amsterdam: Elsevier, 1996): 511–512.

———. (1958). "Quantum Physics and Philosophy: Causality and Complementarity," in J. Kalckar (ed.), *Niels Bohr: Collected Works. Volume 7: Foundations of Quantum Physics II (1933–1958),* (Amsterdam: Elsevier, 1996): 385–394.

Boltzmann, L. (1964). *Lectures on Gas Theory,* translated by S. G. Brush (Berkeley: University of California Press).

Born, M. (1971a), "Comment on Einstein's Letter to Born of March 3, 1947," in M. Born (ed.), *The Born-Einstein Letters* (New York: Walker and Company): 157–158.

———. (1971b), "Comment on Einstein's Letter to Born of April 5, 1948 (containing a translation of Einstein's *Dialectica* paper)," in M. Born (ed.), *The Born-Einstein Letters* (New York: Walker and Company): 173–176.

———. (1971c), "Comment on Einstein's Letter to Born of September 15, 1950," in Born (ed.), *The Born-Einstein Letters* (New York: Walker and Company): 189.

———. (1971d), "Comment on Einstein's Letter to Born of May 12, 1952," in M. Born (ed.), *The Born-Einstein Letters* (New York: Walker and Company): 193.

———.(1971e), "Letter of Pauli to Born of March 3, 1954," in M. Born (ed.), *The Born-Einstein Letters* (New York: Walker and Company): 217–218.

Bostock, D. (2012). *Russell's Logical Atomism* (Oxford: Oxford University Press).

Braithwaite, R. B. (1953). *Scientific Explanation: A Study of the Function of Theory, Probability and Law in Science* (Cambridge: Cambridge University Press).

Brown, H. (2005). *Physical Relativity: Spacetime Structure from a Dynamical Perspective* (Oxford: Oxford University Press).

Bub, J. (2016). *Bananaworld: Quantum Mechanics for Primates* (Oxford: Oxford University Press).

Buchdahl, G. (1964). "Theory Construction: The Work of Norman Robert Campbell," *Isis* **55**: 151–162.

Camilleri, K., and M. Schlosshauer. (2015). "Niels Bohr as Philosopher of Experiment: Does Decoherence Theory Challenge Bohr's Doctrine of Classical Concepts?" *Studies in History and Philosophy of Modern Physics* **49**: 73–83.

Campbell, N. R. (1920). *Physics. The Elements.* (Cambridge: Cambridge University Press).

Carnap, R. (1934). *Logical Syntax of Language,* translated by A. Smeaton (London: Routledge and Kegan Paul, 1937).

———. (1939). "Foundations of Logic and Mathematics," in *International Encyclopedia of Unified Science, Vol. I, Part 1* (Chicago: University of Chicago Press): 139–213.

———. (1950). "Empiricism, Semantics and Ontology," *Revue Internationale de Philosophie* **4**: 20–40, revised and reprinted in *Meaning and Necessity: A Study in Semantics and Modal Logic* (Chicago: University of Chicago Press, second enlarged edition, 1956): 205–221.

———. (1956a). "The Methodological Character of Theoretical Concepts," in H. Feigl and M. Scriven (eds.), *The Foundations of Science and the Concepts of Psychology and Psychoanalysis, Minnesota Studies in the Philosophy of Science, Vol. 1* (Minneapolis: University of Minnesota Press): 38–76.

———. (1956b). *Meaning and Necessity: A Study in Semantics and Modal Logic* (Chicago: University of Chicago Press, second enlarged edition).

———. (1961). "On the Use of Hilbert's ε-operator in Scientific Theories," in Y. Bar-Hillel et al. (eds.), *Essays in the Foundations of Mathematics Dedicated to A. A. Fraenkel* (Jerusalem: Magnus Press of Hebrew University): 156–164.

———. (1963). "Replies and Systematic Expositions," in P. A. Schilpp (ed.), *The Philosophy of Rudolf Carnap* (La Salle IL: Open Court): 859–1013.

———. (1974). *An Introduction to the Philosophy of Science,* Martin Gardner (ed.) (New York: Basic Books).

Chalmers, A. (2011). "Drawing Philosophical Lessons from Perrin's Experiments on Brownian Motion: A Response to van Fraassen," *British Journal for the Philosophy of Science* **62**: 711–732.

Cohen, I. B., and R. E. Schofield. (1978). *Isaac Newton's Letters and Papers on Natural Philosophy,* revised edition (Cambridge, MA: Harvard University Press).

Demopoulos, W. (2003a). "On the Rational Reconstruction of Our Theoretical Knowledge," *British Journal for the Philosophy of Science* **54**: 371–403.

———. (2003b). "Russell's Structuralism and the Absolute Description of the World," in N. Griffin (ed.), *The Cambridge Companion to Russell* (Cambridge: Cambridge University Press): 392–419.

———. (2007). "Carnap on the Rational Reconstruction of Scientific Theories," in M. Friedman and R. Creath (eds.), *The Cambridge Companion to Carnap* (Cambridge: Cambridge University Press): 248–272.

———. (2010). "Effects and Propositions," *Foundations of Physics* **40**: 368–389.

———. (2011a). "Three Views of Theoretical Knowledge," *British Journal for the Philosophy of Science* **62**: 177–205.

———. (2011b). "On Extending 'Empiricism, Semantics and Ontology' to the Realism-Instrumentalism Controversy," *Journal of Philosophy* 108: 647–669.

———. (2013). *Logicism and Its Philosophical Legacy* (Cambridge: Cambridge University Press).

Demopoulos, W., M. Frappier, and J. Bub. (2012). "Poincaré's 'Les conceptions nouvelles de la matière,'" *Studies in History and Philosophy of Modern Physics* **43**: 221–225.

Demopoulos, W., and M. Friedman (1985). "Bertrand Russell's *The Analysis of Matter*: Its Historical Context and Contemporary Interest," *Philosophy of Science* **52**: 621–639.

Dummett, M. (1987). *The Logical Basis of Metaphysics* (Cambridge, MA: Harvard University Press).

Ehlers, J. (1981). "Über den Newtonschen Grenzwert der Einsteinschen Gravitationstheorie." In J. Nitsch, J. Pfarr, and E. W. Stachow (eds.), *Grundlagenprobleme der modernen Physik* (Mannheim: Bibliographisches Institut): 65–84.

Einstein, A. (1905). "On the Movement of Small Particles Suspended in a Stationary Liquid Demanded by the Molecular-Kinetic Theory of Heat," in *Investigations on the Theory of the Brownian Movement,* edited with notes by R. Fürth and translated by A. D. Cowper (New York: Dover, 1956): 1–18.

———. (1906). "On the Theory of Brownian Movement," in *Investigations on the Theory of the Brownian Movement,* edited with notes by R. Fürth and translated by A. D. Cowper (New York: Dover, 1956): 19–35.

———. (1919). "What Is the Theory of Relativity?" *London Times,* November 28.

———. (1921). *Address to the Prussian Academy,* translated by S. Bargmann as "Geometry and Experience," and reprinted in A. Einstein, *Sidelights on Relativity* (New York: E. P. Dutton and Co., 1923): 27–45.

———. (1935). "The Experiment of Einstein, Podolsky and Rosen: A Letter from Albert Einstein," in K. Popper, *The Logic of Scientific Discovery* (New York: Basic Books, 1959): 457–464.

———. (1947). "Letter to Born of March 3, 1947," in M. Born (ed.), *The Born-Einstein Letters* (New York: Walker and Company, 1971a): 157–158.

———. (1948). "Quanten Mechanik und Wirklichkeit," *Dialectica* **2**: 320–324; translated in Born (1971b, pp. 168–173).

———. (1949). "Reply to Criticism," in P. Schilpp (ed.), *Albert Einstein, Philosopher-Scientist: The Library of Living Philosophers, Vol. VII* (Peru, IL: Open Court): 663–688.

———. (1952). "Letter to Born of May 12, 1952," in M. Born (ed.), *The Born-Einstein Letters* (New York: Walker and Company, 1971d): 192.

———. (1953). "Elementare Uberlegungen zur Interpretation der Grundlagen der Quanten-Mechanik," in E. Appleton (ed.), *Scientific Papers Presented to Max Born* (New York: Hafner): 33–40.

Einstein, A., B. Podolsky, and N. Rosen. (1935). "Can Quantum Mechanical Description of Physical Reality be Considered Complete?" *Physical Review* **47**: 777–780.

Feigl, H. (1950). "Existential Hypotheses," *Philosophy of Science* **17**: 192–223.

Freudenthal, H. (1962). "The Main Trends in the Foundations of Geometry in the 19th Century," in E. Nagel et al. (eds.), *Logic, Methodology and Philosophy of Science: Proceedings of the 1960 International Congress* (Stanford, CA: Stanford University Press): 613–621.

Friedman, M. (2012). "Carnap's Philosophical Neutrality between Realism and Instrumentalism," in M. Frappier, D. Brown, and R. DiSalle (eds.), *Analysis and Interpretation in the Exact Sciences: Essays in Honour of William Demopoulos* (Dordrecht: Springer): 95–114.

———. (2017). "Kant's Conception of Causal Necessity and its Legacy," in M. Massimi and A. Breitenbach (eds.), *Kant and the Laws of Nature* (Cambridge: Cambridge University Press): 195–213.

Galavotti, M. (ed.) (1992). *Frank Plumpton Ramsey: Notes on Philosophy, Probability and Mathematics* (Naples: Bibliopolis).

Gleason, A. M. (1957). "Measures on the Closed Subspaces of a Hilbert Space," *Journal of Mathematics and Mechanics* **6**: 885–893.

Glymour, C. (1980). *Theory and Evidence* (Princeton, NJ: Princeton University Press).

———. (2013). "Theoretical Equivalence and the Semantic View of Theories," *Philosophy of Science* **80**: 286–297.

Gupta, A. (2006). *Empiricism and Experience* (Oxford: Oxford University Press).

———. (2009). "Definition," in Edward N. Zalta (ed.), *Stanford Encyclopedia of Philosophy* (Spring 2009 edition).

———. (2011). "Meaning and Misconceptions," in *Truth, Meaning, Experience* (Oxford: Oxford University Press): 164–195.

———. (2013). "The Relationship of Experience to Thought," *The Monist* **96**: 252–294.

Hacking, I. (1983). *Representing and Intervening* (Cambridge: Cambridge University Press).

Hallett, M. (2008). "Reflections on the Purity of Method in Hilbert's *Grundlagen der Geometrie*," in P. Mancosu (ed.) *The Philosophy of Mathematical Practice* (Oxford: Oxford University Press): 198–255.

Halvorson, H. (2012). "What Theories Could Not Be," *Philosophy of Science* **77**: 183–206.

———. (2013). "The Semantic View, If Plausible, Is Syntactic," *Philosophy of Science* **77**: 475–478.

Harrigan, N., and R. Spekkens (2010). "Einstein, Incompleteness, and the Epistemic View of Quantum States," *Foundations of Physics* **40**(2): 125–157.

Hempel, C. G. (1963). "Implications of Carnap's Work for the Philosophy of Science," in P. A. Schilpp (ed.), *The Philosophy of Rudolf Carnap* (La Salle, IL: Open Court): 685–710.

———. (1965). "The Theoretician's Dilemma: A Study in the Logic of Theory Construction," in *Aspects of Scientific Explanation* (New York: Free Press, 1965): 173–226.

———. (1970). "On the 'Standard Conception' of Scientific Theories," in M. Radner and S. Winokur (eds.), *Analyses of Theories and Methods of Physics and Psychology, Minnesota Studies in the Philosophy of Science, Vol. 4* (Minneapolis: University of Minnesota Press): 142–163.

Howard, D. (1985). "Einstein on Locality and Separability," *Studies in History and Philosophy of Science A* **16**(3): 171–201.

———. (1989). "Holism, Separability, and the Metaphysical Implications of the Bell Experiments," in J. Cushing and E. McMullin (eds.), *Philosophical Consequences of Quantum Theory* (South Bend, IN: University of Notre Dame Press): 224–253.

Janssen, M. (2002). "COI Stories: Explanation and Evidence in the History of Science," *Perspectives on Science* **10**: 457–522.

Kochen, S., and E. Specker (1967). "The Problem of Hidden Variables in Quantum Mechanics," *Journal of Mathematics and Mechanics* **17**: 59–87.

Krips, H. (1986). "Atomism, Poincaré and Planck," *Studies in History and Philosophy of Science* **17**: 43–63.

Landsman, N. P. (2008). "Macroscopic Observables and the Born Rule, I. Long Run Frequencies," *Reviews in Mathematical Physics* **20**(10): 173–190.

Laudan, L. (1981). *Science and Hypothesis* (Dordrecht: Reidel).

Lewis, D. (1970). "How to Define Theoretical Terms," *Journal of Philosophy* **67**: 427–445.

———. (1983). "New Work for a Theory of Universals," *Australasian Journal of Philosophy* **61**: 343–377.

———. (1984). "Putnam's Paradox," *Australasian Journal of Philosophy* **62**: 221–236.

Maddy, P. (2007). *Second Philosophy: A Naturalistic Method* (Oxford: Oxford University Press).

Malament, D. (1986). "Newtonian Gravity, Limits, and the Geometry of Space," in R. Colodny (ed.), *From Quarks to Quasars: Philosophical Problems in Modern Physics* (Pittsburgh, PA: University of Pittsburgh Press): 181–202.

Maudlin, T. (2014). "Reply to Comment on 'What Bell Did,'" *Journal of Physics A* **47**: 424012.

Maxwell, J. C. (1875). "Atom," in *Encyclopaedia Britannica, Ninth Edition, Vol. 3* (Edinburg: Adam and Charles Black).

Merrill, G. (1980). "The Model-Theoretic Argument against Realism," *Philosophy of Science* **47**: 69–81.

Millikan, R. (1917). *The Electron, Its Isolation and Measurement and the Determination of Some of Its Properties* (Chicago: University of Chicago Press).

———. (1924). "The Electron and the Light Quant from the Experimental Point of View," *Nobel Lecture,* May 23, 1924.

Myrvold, W. (2003). "On Some Early Objections to Bohm's Theory," *International Studies in the Philosophy of Science* **1**: 7–24.

———. (2016). "Lessons of Bell's Theorem: Nonlocality, Yes; Action at a Distance, Not Necessarily," in M. Bell and S. Gao (eds.), *Quantum Nonlocality and Reality* (Cambridge: Cambridge University Press): 238–260.

Nagel, E. (1961). *The Structure of Science* (New York: Harcourt, Brace and World).

Nernst, W. (1893). *Theoretische Chemie vom Standpunkte der Avogadroschen Regel und der Thermodynamik.* 1st edition (Stuttgart: Ferdinand Enke).

Newman, M. H. A. (1928). "Mr Russell's Causal Theory of Perception," *Mind* **37**: 137–148.

Newton I. (1687). *The Principia: Mathematical Principles of Natural Philosophy.* In I. B. Cohen, A. W. Whitman, and J. Budenz (eds.), *The Principia: Mathematical Principles of Natural Philosophy,* trans. I. B. Cohen & A. Whitman (Los Angeles: University of California Press, 1999).

Norsen, T. (2009). "Local Causality and Completeness: Bell vs. Jarrett," *Foundations of Physics* **39**: 273–294.

Norton, J. (2000). "How We Know about Electrons," in R. Nola and H. Sankey (eds.), *After Popper, Kuhn and Feyerabend* (Dordrecht: Kluwer): 67–97.

Ostwald, W. (1909). *Grundriss der allgemeinen Chemie,* 4th ed. (Leipzig: Akad. Verl.- Ges).

Pauli, W. (1971). "Letter to Born of March 3, 1954," in M. Born (ed.), *The Born-Einstein Letters* (New York: Walker and Company): 217–219.

Perrin, J. (1910). *Brownian Movement and Molecular Reality,* translated from the *Annales de Chime et de Physique, 8th Series,* September 1909, by F. Soddy (London: Taylor and Francis).

———. (1916). *Atoms,* authorized translation by D. Ll. Hammock (New York: Van Nostrand).

———. (1926). "Discontinuous Structure of Matter," *Nobel Lecture,* December 11, 1926.

Pitowsky, I. (1989). *Quantum Probability–Quantum Logic* (Berlin: Springer-Verlag).

———. (1994). "George Boole's 'Conditions of Experience' and the Quantum Puzzle," *British Journal for the Philosophy of Science* **45**: 95–125.

———. (2006). "Quantum Mechanics as a Theory of Probability," in W. Demopoulos and I. Pitowsky (eds.), *Physical Theory and Its Interpretation. The Western Ontario Series in Philosophy of Science, Vol. 72* (Dordrecht: Springer): 213–240.

Poincaré, H. (1902). *Science and Hypothesis,* in *The Foundations of Science* (Lancaster, PA: Science Press, authorized 1913 translation of George Bruce Halsted); see also the alternative translation, a Dover (New York, 1952) reprint.

———. (1905). *The Value of Science,* authorized 1913 translation of George Bruce Halsted. (New York: Dover, 1958).

———. (1906). "Letter to the Editor," *Mind* **15**: 141–143.

———. (1912a). "Les conceptions nouvelles de la matière," *Foi et Vie* **15**: 185–191. Citations of this essay are to the translation that appears in Demopoulos, Frappier, and Bub (2012).

———. (1912b). "The Relations between Matter and Ether,' in H. Poincaré, *Mathematics and Science: Last Essays,* translated by John W. Bolduc (New York: Dover, 1963): 89–101.

Popescu, S., and D. Rohrlich. (1994). "Quantum Nonlocality as an Axiom," *Foundations of Physics* **24**(3): 379–385.

Psillos, S. (1999). *Scientific Realism: How Science Tracks Truth* (London: Routledge).

———. (2006). "Ramsey's Ramsey Sentence," in M. Galavotti (ed.), *Cambridge and Vienna: Frank P. Ramsey and the Vienna Circle* (Dordrecht: Springer): 77–100.

———. (2011). "Moving Molecules above the Scientific Horizon: On Perrin's Case for Realism," *Journal for General Philosophy of Science.* Published online, September 18.

Putnam, H. (1962). "What Theories Are Not," in E. Nagel et al. (eds.), *Logic, Methodology and Philosophy of Science: Proceedings of the 1960 International Congress* (Stanford, CA: Stanford University Press): 240–251.

———. (1976). "Realism and Reason," *Proceedings and Addresses of the American Philosophical Association* **50**: 483–498.

———. (2012). "A Theorem of Craig's about Ramsey Sentences," in M. DeCaro and D. Macarthur (eds.), *Philosophy in an Age of Science* (Cambridge, MA: Harvard University Press): 277–279.

Quine, W. V. O. (1951). "Two Dogmas of Empiricism," reprinted with abridgements and expansions in *From a Logical Point of View: Logico-Philosophical Essays* (Cambridge, MA: Harvard University Press, second edition, 1980): 20–46.

Ramsey, F. P. (1925), "Universals," in R. B. Braithwaite (ed.), *The Foundations of Mathematics and Other Logical Essays by Frank Plumpton Ramsey* (Paterson, NJ: Littlefield and Adams, 1931): 112–134.

———. (1929), "Theories," in R. B. Braithwaite (ed.), *The Foundations of Mathematics and Other Logical Essays by Frank Plumpton Ramsey* (Paterson, NJ: Littlefield and Adams, 1931): 212–236.

Reichenbach, H. (1920). *The Theory of Relativity and A Priori Knowledge,* translated and edited with an introduction by M. Reichenbach (Berkeley and Los Angeles: University of California Press, 1965).

Renn, J. (2005). "Einstein's Invention of Brownian Motion," *Annalen der Physik* (Leipzig) **14**, Supplement: 23–37.

Russell, B. (1905). "Review of Science and Hypothesis," in *Philosophical Essays* (New York: Simon and Shuster, 1966): pp. 70–78.

———. (1927). *The Analysis of Matter* (New York: Dover, 1954).

———. (1968). *Autobiography.* Vol. 2, 1914–1944 (London: Allen & Unwin).

Schilpp, P. A. (ed.) (1963). *The Philosophy of Rudolf Carnap* (La Salle IL: Open Court).

Schlick, M. (1918). *General Theory of Knowledge,* translated by A. E. Blumberg (La Salle, IL: Open Court, 1985).

Schrödinger, E. (1935). "Letter to Bohr of October 13th, 1935," in J. Kalckar (ed.), *Niels Bohr: Collected Works. Vol. 7: Foundations of Quantum Physics II (1933–1958)* (Amsterdam: Elsevier, 1996): 507–509.

Shimony, A. (1993). *Search for a Naturalistic Worldview, Vol. 2: Natural Science and Metaphysics* (Cambridge: Cambridge University Press).

Smith, G. (2001). "J. J. Thomson and the Electron, 1897–1899," in J. Z. Buchwald and A. Warwick (eds.), *Histories of the Electron: The Birth of Microphysics* (Cambridge, MA: MIT Press, 2001): 21–76.

———. (2002). "The Methodology of the *Principia,*" in I. B. Cohen and G. E. Smith (eds.), *The Cambridge Companion to Newton* (Cambridge: Cambridge University Press, 2002): 138–173.

Smith, G., and R. Seth. (2020). *Brownian Motion and Molecular Reality: A Study in Theory-Mediated Measurement* (Oxford: Oxford University Press).

Solomon, G. (1989). "An Addendum to Demopoulos and Friedman (1985)," *Philosophy of Science* **56**: 497–501.

Stairs, A. (2011). "A Loose and Separate Certainty: Caves, Fuchs and Schack on Quantum Probability One," *Studies in History and Philosophy of Modern Physics* **42**(3): 158–166.

Stanford, P. K. (2009). "Scientific Realism, the Atomic Theory, and the Catch-All Hypothesis: Can We Test Fundamental Theories against All Serious Alternatives?" *British Journal for the Philosophy of Science* **60**: 253–269.

Stein, H. (1967). "Newtonian Space-Time," *Texas Quarterly* **10**: 174–200.

Teller, P. (1989). "Relativity, Relational Holism, and the Bell Inequalities," in J. Cushing and E. McMullin (eds.), *Philosophical Lessons from Quantum Theory* (South Bend, IN: Notre Dame Press): 208–223.

Thomson, J. J. (1906). "Carriers of Negative Electricity," *Nobel Lecture,* December 11, 1906.

———. (1897). "Cathode Rays," Discourse in Physical Science, Royal Institution. Friday April 30, 1897, in W. L. Bragg and G. Porter, (eds.), *Royal Institution Library of Science, Physical Sciences, Vol. 5* (London: Applied Sciences Publishers): 36–49.

Townsend, J. (1900). "On the Diffusion of Ions into Gases," *Philosophical Transactions of the Royal Society* **A 193**: 129–158.

Van Benthem, J. (1978). "Ramsey Eliminability," *Studia Logica* **37**: 321–336.

Van Fraassen, B. C. (1980). *The Scientific Image* (Oxford: Oxford University Press).

———. (2009). "The Perils of Perrin, in the Hands of Philosophers," *Philosophical Studies* **143**: 5–24.

Van 't Hoff, J. (1901). "Osmotic Pressure and Chemical Equilibrium," *Nobel Lecture,* December 13, 1901.

Von Plato, J. (1994). *Creating Modern Probability* (Cambridge: Cambridge University Press).

Von Smoluchowski, M. (1906). "Zur kinetischen Theorie der Brownschen Molekularbewegung und der Suspensionen," *Annalen der Physik* **21**: 756–780.

Werner, R. (2018). "What Maudlin Replied To." https://arxiv.org/abs/1411.2120.

Winnie, J. (1967). "The Implicit Definition of Theoretical Terms," *British Journal for the Philosophy of Science* **18**: 223–229.

———. (1970). "Theoretical Analyticity," in R. Cohen and M. Wartofsky (eds.), *Boston Studies in the Philosophy of Science, Vol. VIII* (Dordrecht and Boston: Reidel): 289–305.

Worrall, J. (2012). "Miracles and Structural Realism," in E. Landry and D. Rickles (eds.), *Structural Realism* (New York: Springer): 77–95.

ACKNOWLEDGMENTS

It is a bittersweet moment to celebrate the life of William ("Bill") Demopoulos, husband, father, philosopher, and close friend. We now celebrate Bill's final philosophical book, *On Theories,* which had been all but finished when he left us for good. I was able to contribute to the finishing touches on this book in my role as editor, but there are many more people who have contributed to the richness of his life and works.

The dedication of his last published book, *Logicism and its Philosophical Legacy,* reads "*to Rai, Billy, and Alexis, for whom my gratitude is beyond expression.*" It is true that his family, wife and two sons, were (and in an important sense still are) at the very center of his life, just as he was (and continues to be) at the center of theirs.

Bill's life was essentially that of a philosopher. He was involved with the publication of eight books, including *On Theories,* which, we believe, represents the culmination of his philosophical life's work. It is also worth emphasizing that *Analysis and Interpretation in the Exact Sciences: Essays in Honour of William Demopoulos,* appeared as a *Festschrift,* which shows all that he has done for his students.

Bill cultivated close friendships with some of the most important figures in logic, philosophy of science, and cognitive science. Central to his philosophical development was the relationship with Jeffrey Bub, one of the leading philosophers of quantum theory in the world. Bub, in fact, had brought Bill to the University of Western Ontario in the first place, and the two quickly became partners in the Bub-Demopoulos version of quantum logic—whether understood as genuine logic or rather an axiomatization of quantum theory. It is fitting, therefore, that the last chapter of *On Theories* develops an extraordinarily nuanced treatment of "quantum reality" within a novel reconstruction of Bohrian complementarity.

Close friendships with some of the most prominent philosophers in the world continued with Michael Dummett leading the way. Bill had met Dummett at Harvard and then seen him often in the UK, including visits from Bill and his family. Similarly, Bill had met Hilary Putnam also at Harvard, and the two became close as well, especially in connection with quantum theory. To take another example, moreover, Bill befriended David Malament, an eminent philosopher of science. I will always remember that the two of them loved to walk and talk in the woods near Bill's home discoursing on logic, mathematics, and the foundations of physics.

Among some of Bill's more recent colleagues at Western Ontario, Robert DiSalle became Bill's closest colleague and philosophical friend, sharing, in particular, the development of philosophy of science in a more general setting. Zenon Pylyshyn, a leading philosopher of cognitive science, worked closely with Bill in writing joint papers and organizing conferences in the field. Finally, it is important not to forget Itamar Pitowsky, who came from Hebrew University to Western Ontario as a doctoral student. He later became a regular Visiting Scholar and teacher at Western for a number of years. Although Itamar died too young, his extraordinary mathematical and philosophical abilities served Bill's work on quantum theory very well, as one can see from the Bibliography in *On Theories*.

Returning now to close colleagues throughout the world, with no or little connection with Western Ontario, one can name Harvey Brown at Oxford, Anil Gupta at Pittsburgh, Crispin Wright at NYU (and Stirling), Thomas Uebel at Manchester.

Among those who were most active in engaging with Bill during the writing of *On Theories* are Yemima Ben-Menahem at the Hebrew

University, Jeffrey Bub at Maryland, and, of course, Robert DiSalle at Western. One should also add to this list some of Bill's currently most active former students, such as Mélanie Frappier (University of King's College) and Gregory Lavers (Concordia University).

I am sure that I have missed some other close friends and colleagues throughout the world. I am nevertheless pleased, despite the bittersweet moment of finally publishing *On Theories,* that we have now honored this moment appropriately and well.

Michael Friedman

Editor's postscript: I want to thank Adam Zweber for his work on the index.

Portions of chapter 1 were first published as "Logical empiricist reconstructions of theoretical knowledge" in *Logic, Methodology and Philosophy of Science: Proceedings of the 15th International Congress (Helsinki)* edited by Hannes Leitgeb et al., and published by College Publications in 2017.

INDEX

acceleration field property, xvi–xx.
See also Stein, Howard

action-at-a-distance, xiv, 149, 153.
See also principle of local a *Z Analysis
of Matter, The* (Russell), 39–43,
46, 48

atomism, metaphysical, 10, 70, 99.
See also corpuscularean philosophy

Avogadro's number, 7, 66, 71–73, 108,
205n3, 215n28

Bell, John Stewart, 17; analysis of local
causality, 158–167, 181, 195–196,
200–204, 223n40; and causal structure, 158–159, 179–180; theorem of,
154–158, 162–165, 169, 171–172,
179–181, 195–196, 200–202, 223n40

Bell's theorem. See under Bell, John
Stewart

Ben-Menahem, Yemima, 194, 221n16,
224n50

Bohm, David, 220n12, 221n14; and
completeness, 140–142

Bohr, Niels: and complementarity, 126,
128, 139–140, 146, 169, 187–188; and

completeness, 126, 128, 139–140, 143,
154, 169–172, 195, 203; and disturbance interpretations, 146–147;
early theory of atomic structure,
xxiii; and non-commutativity,
127–128; and primacy of classical
concepts, 15, 17, 121–128, 132, 135,
139, 187, 199; and recoverability
of classical mechanics, 135–139;
response to EPR, 148–151, 154,
169–172, 195, 219n3, 221n21; and
special relativity, 130; and visualizability, 135

Bohr / Schrödinger correspondence,
123–130, 218–219n3; 219n4

bootstrapping. See Glymour,
Clark

Born, Max, 143–146, 222n25. See also
Born / Einstein correspondence;
quantum mechanics, statistical
interpretations of

Born / Einstein correspondence,
143–146, 222n29

Brahe, Tycho, 113–114

Brown, Harvey, 208n13, 223n48

Brownian motion: as "connecting link," 10, 81–83, 87, 89, 92; Einstein's theoretical analysis of, 64–67; fine structure of, 75–76, 78; and Poincaré, 101, 107. *See also* Einstein, Albert; granular experiments; Perrin, Jean

Bub, Jeffrey, 150, 188–189, 192–193; and no-signaling condition, 150, 223n41; translation of Poincaré, 215n25, 217nn6–7

Camilleri, Kristian and Schlosshauer, Maximilian, 135–139, 199

Campbell, Norman, 42, 47–48

Carnap, Rudolf, viii–xii, 4, 9, 56, 206–207n9, 207n10; "Empiricism, Semantics, and Ontology," viii, xi, 12; linguistic frameworks, ix, x, 10; metaphysical neutrality, xi, 12, 61–62, 212n34. *See also* Ramsey-sentence reconstruction

Carnap sentences, 21–23, 28–30. *See also* Carnap, Rudolf: Ramsey-sentence reconstruction; theoretical analyticity

cathode rays, 13, 93–98, 216n33. *See also* Thomson, Joseph John

causal theory of perception, 39, 117, 120. *See also* Russell, Bertrand; structuralism

coherence justification of probability axioms, 174–177

complementarity, 126, 187; ambiguity of, 139, 169; and disturbance inter-pretations of quantum mechanics, 169; and freedom of experimenter, 127–128, 140, 222n37. *See also* Bohr, Niels

conditional statistical independence. *See* factorizability

connecting link. *See under* Perrin, Jean

consilience, 110–112, 217nn9–10. *See also* robustness

constructive empiricism: central tenet of, 53–54, 57–61; explication of empirical adequacy, 53–56; model-theoretic objection to, 57–63. *See also* empirical adequacy; Van Fraassen, Bas

constructive versus principle theories, 16–17; and molecular hypothesis, 64–65, 86, 114; and quantum mechanics, 17, 142, 174, 183, 185; and special relativity, 17, 171, 174, 183, 208n13

corpuscularean philosophy, 7, 10, 70, 205–206n4. *See also* atomism, metaphysical

correspondence rules, 2, 11, 19–20, 33, 35, 56, 207n9; in Carnap's Ramsey-sentence reconstructions, 21, 23, 29–30. *See also* Campbell, Norman; Carnap, Rudolf; Ramsey sentence

Craig transcription, 52, 209n6

dictionary, Campbell's notion of. *See* Campbell, Norman

Dummett, Michael, 190, 222n23

effective equivalence of classical and quantum descriptions, 138

Ehlers, Jürgen, 132–134. *See also* Malament, David

Einstein, Albert, ix, 154, 172; analysis of Brownian motion, x, 65–67, 86, 89, 207n10; and constructive versus principle theories, 16–17, 173, 208n13; differences with EPR, 151, 221n16, 222n24; field theoretic program, 152, 162; his real factual situations as truth-value assignments, 158, 164, 178–180, 185; his realism, 149–150, 161, 183–184, 201; interpretation of Bohr, 148–151, 154, 169–170; objection to disturbance interpretations, 144; principle of local action, 151–154; and statistical interpretations, 144–146

Einstein-Podolsky-Rosen (EPR), 17; argument of, 143–144, 195–196; Bell's response to (*see* Bell, John Stewart); Bohr's response to, 136, 147–151, 154, 169–171; criterion of reality, 221n19; differences with Einstein, 151, 221n16; primary target of, 144–146, 150–151

Einstein's commitment to mutual independence, 148–150, 152, 161–162, 165, 169, 172, 178, 182–183

Einstein's local realism. *See* Bell, John Stewart: analysis of local causality

Einstein's realism, 149–150, 161, 183–184, 201. *See also* Bell, John Stewart: analysis of local causality

electromagnetic constitution of matter, 16, 173, 183, 223n47

electron, charge of, 6, 94–98, 122. *See also* Millikan, Robert; Thomson, Joseph John

empirical adequacy, 53–62, 69, 90, 212n32

empirical well-foundedness, xiv, 5, 11, 62, 88–89, 108–109; versus Van Fraassen's empirical grounding, 54, 90–91, 215n28

"Empiricism, Semantics and Ontology" (Carnap), viii, xi, 12, 206nn6–7

epistemic interpretation of the wave function, 151, 220n9

epsilon operator, 23, 62. *See also* Carnap, Rudolf; Hilbert, David

equipartition of energy, xxiii, 64, 89

ether, xiv, 87–88, 125

evidentiary frameworks versus theoretical frameworks, 218n2; and Bohr, 121–123, 128–129, 135, 138–139, 187–188, 199; and general relativity, 132–134; and Newtonian methodology, 188, 198–200; and probability in quantum mechanics, 168–169

explicit definition, 20–24. *See also* implicit definition

factorizability, 155, 161–164

factual content of a theory, xi, 20–21, 32. *See also* Ramsey sentence

finitism, 31, 68–69. *See also* Hilbert, David

Franklin, Benjamin, 95

Frege, Gottlob, viii, 36, 190

Friedman, Michael, 39–40, 131–132

Gay-Lussac's law, 105–106

general relativity, 17, 180; as analogous to quantum mechanics, 188; as disanalogous to quantum mechanics, 129, 131–134; evidentiary framework of, 131–134, 219n5; and Newtonian methodology, xx, xxiv, 153, 187, 198

geometry: in evidentiary versus theoretical frameworks, 131–134, 219n5; and Russell, 43–44

Gleason's theorem, 168, 179, 194–195, 202–203; and Bohr, 170–172. *See also* quantum mechanics, noncontextuality of

Glymour, Clark, xxii, 215n27

Gödel's theorems, 68–69

gold leaf experiments, 95. *See also* Hertz, Heinrich; Thomson, Joseph John

Gouy, Louis-Georges, 95

granular and molecular parameters, 83, 91. *See also* Perrin, Jean; *specific parameters*

granular energy, 73, 78, 83, 91. *See also* Perrin, Jean

granular experiments, 73–76, 78–80, 83–88

Hacking, Ian, 207–208n11

Halvorson, Hans, 211–212n32

Hertz, Heinrich, 95, 97

hidden variables. *See* Bohm, David

Hilbert, David, xi–xiii, 2, 19, 23, 35–36, 62, 68

Howard, Don, 162, 221n16, 222n24

Hume's Principle, vii–ix; criteria of identity in mathematics versus physics, ix. *See also* mathematical versus empirical theories

hypothetico-deductive reasoning, x, 3, 70, 75, 93, 109, 113. *See also* method of hypothesis

implicit definition, 19, 22. *See also* Carnap sentences; Hilbert, David

inference to the best explanation, 3, 70, 92–93, 109, 113. *See also* method of hypothesis

irreducibly statistical theory, 178–183, 202–203

Kant, Immanuel, 68, 131–132

Kochen and Specker's Theorem, 141, 220n12; compared with Bell's theorem, 157, 172; differences with Bell's/Demopolous's approach, 167, 189, 191, 200–201, 204; Pitowsky's formulation of, 156

Landsman, Klaas, 137–139, 219nn7–8

Laudan, Larry, 111–112, 217n11

Lewis, David, 4; and logical empiricism, 28–32; and natural properties, 37–39, 49; response to Newman's objection, 210n14; response to Winnie, 24–28

local causality, 155, 158, 161–164, 201–203. *See also* Bell, John Stewart; nonlocality; relativistic causality

logical empiricism and Hilbertian axiomatization, xi–xiii. *See also*

Hilbert, David; Ramsey, Frank Plumpton; Russell, Bertrand; structuralism: structuralist thesis logicism, viii–ix, 43, 52–53. *See also* Russell, Bertrand

Lorentz, Hendrik, 173, 223n47. *See also* special relativity

Mach, Ernst, 115–116
Maddy, Penelope, ix–x, 207n10
Malament, David, 132, 134, 219n6. *See also* Ehlers, Jürgen
mathematical versus empirical theories, ix, xi–xii, xxiv, 2, 19–20, 31–32, 35–38, 211n32
Maudlin, Tim, 223n40
Maxwell-Boltzmann distribution law, xxii–xxiii, 64
Maxwell's laws of electro-magnetism, 16, 207n9, 208n13. *See also* electromagnetic constitution of matter
Mercury, anomaly of, xxi–xxii; and general relativity, xx–xxii
method of hypothesis, 3–5, 10, 77, 108–110, 114. *See also* hypothetico-deductive reasoning; inference to the best explanation
Millikan, Robert, 97–98, 205n2, 207n11
Minkowski spacetime, 51; causal structure of, 158–159, 179–180, 200; and constructive versus principle theories, 16–17, 172–174, 183, 208n13; as disanalogous to algebraic structure of quantum mechanics, 129–130

mixed vocabulary, xii, 2, 18, 25, 27, 29. *See also* correspondence rules
model-theoretic argument, 32–34; Lewis's reply to, 37–39; and Newman's objection, 40; significance of, 35–36. *See also* Putnam, Hilary
molecular diameter, 66, 122, 213n7
molecular hypothesis, definition of, 64. *See also* molecular-kinetic theory
molecular-kinetic theory: limited validity of, 88–89; versus molecular hypothesis, 64–65, 88–89; principle and constructive parts of, 64, 86, 208n13, 212n2. *See also* Perrin, Jean
multiple realizability, 24, 32, 34, 36. *See also* Lewis, David
mutual independence, 148–150, 152, 161–162, 165, 169, 172, 178; and no-signaling condition, 182–183. *See also under* Bub, Jeffrey

natural properties. *See under* Lewis, David; Ramsey, Frank Plumpton
Neptune, discovery of, xix—xxi
Newman's objection, 39–41, 210n18
Newton, Isaac, 70, 101, 181; importance as a methodologist, xiii, xvi–xviii, xxi, xxiii, 5, 187, 197, 204; *Principia*, xiii–xiv, xvi, 198
Newtonian gravity, xiv–xxiv, xxiii–xiv, 187, 200, 204; acceleration field property, xvi–xx; as an instrument, xxi–xxii; as limit of general relativity, 132–134; quasi-closed systems in, 153, 196–198; Trautman-Cartan formulation of, 132–134

Newtonian methodology, xiii–xxiv, 197–198, 204; Rules for the Study of Natural Philosophy, xiii–xiv, xvii–xviii, xxi, xxiii–xxiv, 197–198, 204; and theory-mediated measurement, xiii, xv–xvii, xxii, 5, 153, 187, 196–200. *See also* Smith, George; theory-mediated measurements

nonlocality, 180–181. *See also* local causality; Popescu, Sandu and Rohrlich, Daniel; principle of local action

no-signaling condition, 150; as extension of mutual independence, 182–183; and parameter independence, 164–166. *See also* Bub, Jeffrey

osmotic pressure, 79, 104–106. *See also* Perrin, Jean; Poincaré, Henri; van 't Hoff, Jacobus

Ostwald, Wilhelm, 70, 216n31

outcome independence, 164–166

O-vocabulary. *See* Lewis, David

parameter independence, 164–166

partial interpretation, definition of, 18–19

periodic phenomena, 101–102. *See also* Poincaré, Henri

Perrin, Jean, ix–x, 13–15, 65, 67–70, 93–95; connecting link, 10–11, 78–83, 87–89, 92; and law, 122; use of theory-mediated measurements, xxi–xxii, 7–8, 87–88, 90–91, 109, 187, 205n3; Van Fraassen's reading of, xxii, 80–81, 84, 90–91, 214n17, 214n19

"Physics Says" (Ramsey), 40

Pitowsky, Itamar, 155–158, 163, 174–176, 179, 194, 201, 222n31, 222nn33–34

Podolsky, Boris. *See* Einstein-Podolsky-Rosen

Poincaré, Henri, vii, 13, 88, 216–217n6; and consilience-based arguments, 110–113, 217nn9–10; indifferent hypotheses, 107–109, 216n5; Russell's reading of, 116–120, 218n17; *Science and Hypothesis*, 100, 116; and scientific realism, 112–115; and true relations, 100–110, 120, 122–123, 217n9

Popescu, Sandu and Rohrlich, Daniel, 180–181, 224n49

Popper, Karl, 144–145, 151, 222n24

prediction versus measurement, 5–6, 8, 112, 199. *See also* robustness; theory-mediated measurements

primacy of classical concepts, 15, 17, 121–122, 187; in quantum mechanics versus relativity, 132–135; and recoverability of classical mechanics, 135–139; Schrödinger's objection to, 123–127. *See also* Bohr, Niels; Bohr / Schrödinger correspondence

Principia (Newton), xiii–xiv, xvi, 198

Principia Mathematica (Russell), 41–44, 47

principle of local action, 151–154. *See also* Bell, John Stewart: analysis of local causality; Bell, John Stewart: and causal structure

probability, classical versus quantum mechanical, 184–185; and Bell's theorem, 157–158, 169, 179–180, 201; and Gleason's theorem, 168–169, 179; and irreducibly statistical theories, 178; justification of principles, 174–177; and Popescu and Rohrlich, 180–181; and principle theories, 173–174; relative frequencies in, 168, 177–178, 203, 223n44. *See also* Pitowsky, Itamar; quantum mechanics, noncontextuality of

probability, epistemic concept of, 174–179

Psillos, Stathis, 49–50, 74–75, 210n14

Putnam, Hilary, 4, 32–40, 210n18; and quantum logic, 188. *See also* model-theoretic argument

quantum gambles, 175–178

quantum indeterminacy, 139, 169, 180–181

quantum mechanical representation of measuring instruments, 138

quantum mechanics, disturbance interpretations of, 144–147

quantum mechanics, noncontextuality of, 175, 191; empirical justification of, 177, 192, 202; and Gleason's theorem, 179, 193–194, 202–203. *See also* probability, classical versus quantum mechanical

quantum mechanics, observables in, 128–131, 137–138, 141–142, 145, 167–168

quantum mechanics, orthodox interpretations of, 15, 148, 154, 184

quantum mechanics, statistical interpretations of, 143–146

quasi-closed systems, 152–154, 196–198, 201. *See also* Newtonian methodology

Quine, Willard Van Orman, 205n4

Ramsey, Frank Plumpton, viii, 4, 21; and empirical character of physics, 42–43, 52; and natural [P]roperties, 49–51; "Physics Says," 40–47; "Theories," 41–42, 47–49; and truth in physics, 46–49, 51, 53; "Universals," 49–51

Ramsey sentence, 36, 38, 52, 209n6; and *Analysis of Matter*, 40, 47–49; arithmetical interpretation of, 32; Carnap's use of, xi–xii, 20–23, 28–32, 62, 209n12; Lewis's use of, 28–29; matrix of, xii, 21, 23, 208n5

Ramsey-sentence reconstruction, xi– xii, 20–25, 29–30, 32

real factual situations, 154, 172, 179, 183; and Bohr's reply to EPR, 169–170; as truth-value assignments, 158, 164, 178, 180, 185. *See also* Einstein, Albert

realism: and Demopoulos's interpretation of quantum mechanics, 183–184; and Einstein (*see* Einstein's realism); entity, 207–208n11; and instrumentalism, ix–xi, 12–17, 24, 62, 116, 212n34; metaphysical, 35; and Poincaré, 113–116; scientific, 13–14, 57–59, 112–115, 149

realist extensions of quantum mechanics, 14–15, 141–142

recoverability of classical mechanics, 132–136

Reichenbach, Hans, 215n28

relativistic causality, 180–181, 196, 200–203

relativity principles, 16–17. *See also* special relativity; general relativity

robustness, 7–9, 62–63, 91–92, 94, 108–112. *See also* theory-mediated measurements

Rosen, Nathan. *See* Einstein-Podolsky-Rosen

Russell, Bertrand: *Analysis of Matter, The,* 39–43, 46, 48; main differences with Ramsey, 51–53; and Newman's objection, 39–40; *Principia Mathematica,* 41–42, 44, 47; reading of Poincaré, 116–120; and structuralism, 39–40, 46, 49–53

Schrödinger, Erwin, 124–128, 146, 218–219n3, 221n16. *See also* Bohr / Schrödinger correspondence

Science and Hypothesis (Poincaré), 100, 116

Scientific Image, The (Van Fraassen), 54–60

semantic view of theories. *See* constructive empiricism; Van Fraassen, Bas

separability of properties, 162

Smith, George, xiii, 205n1

special relativity: and Bell's theorem, 179–180; as disanalogous to quantum mechanics, 129–131; as a principle theory, 16–17, 171–174, 183, 208n13. *See also* constructive versus principle theories; Minkowski spacetime

Stein, Howard, xvi

stochastic hidden variable theories, 165–166, 223n42

Stokes's Law, 6, 97–98, 122

structuralism: as distinct from Poincaré's view, 114, 118–120; Newman's objection, 39–40; Russell's objection, 39–40, 46, 49, 53, 118–120, 218n15; scientific structural realism, 114; structuralist thesis, xii–xiii, 20–21, 27

superquantum theory / correlations, 181–182

surface / operational level of measurable parameters, 150, 159–161, 165, 182, 200, 204. *See also* Bell, John Stewart: analysis of local causality

syntactic view of theories, 3, 30–32, 55. *See also* Halvorson, Hans; mathematical versus empirical theories

Teller, Paul, 162

theoretical analyticity, 21–22. *See also* Carnap sentences

"Theories" (Ramsey), 41–42, 47–49

theory-mediated measurements: and logical empiricism, 11, 62–63; Perrin's use of, xxii, 5–6, 7–8, 8, 15, 112, 199; Poincaré's use of, 104–110; robustness of, 7, 9, 94, 109–113; Thomson's use of, 6–7, 13–15, 94, 99;

Townsend's use of, 5–7. *See also* Newtonian methodology

Thomson, Joseph John, 93–99, 215n30, 216nn31–33; and theory-mediated measurements, 6–7, 13–15, 94, 99; use of Stokes's law, 6, 97–98, 122

Townsend, John, 5–7, 94

true relations. *See under* Poincaré, Henri

truth: approximate, 12, 97, 114–115, 149; versus empirical adequacy, 54, 56–63; as predictive success, 48–49, 90, 114–115; Russell's conception of, 41–43, 46–48, 51, 53; versus satisfiability, 35–36

T-vocabulary. *See* Lewis, David

Tycho. *See* Brahe, Tycho

underlying parameters, 160–165, 182. *See also* Bell, John Stewart: analysis of local causality

"Universals" (Ramsey), 49–51

Van Benthem, Johann, 33

Van Fraassen, Bas: and assumptions of Perrin, xxii, 80–81, 84; and constructive empiricism (*see* constructive empiricism); notion of empirical grounding, 90–91; *Scientific Image, The,* 54–60

van 't Hoff, Jacobus, 79–81, 87, 105–106

vertical concentration experiments. *See* granular experiments

visualizability, 135. *See also* Bohr, Niels

Von Nägeli, Karl, 67, 213n8

Von Neumann, John, 141, 220n10, 222n32; formulation of quantum mechanics, 14, 17

water droplet argument, 97–98

Weyl-Minkowski theorem, 222n34

Whewell, William, 110–112, 217n9

Wilson, Charles Thomson Rees, 97

Wilson, Harold Albert, 97

Winnie, John, 4, 22, 24–28, 32, 34